Advanced Sciences and Technologies for Security Applications

Indexed by SCOPUS

The series Advanced Sciences and Technologies for Security Applications comprises interdisciplinary research covering the theory, foundations and domain-specific topics pertaining to security. Publications within the series are peer-reviewed monographs and edited works in the areas of:

- biological and chemical threat recognition and detection (e.g., biosensors, aerosols, forensics)
- crisis and disaster management
- terrorism
- cyber security and secure information systems (e.g., encryption, optical and photonic systems)
- traditional and non-traditional security
- energy, food and resource security
- economic security and securitization (including associated infrastructures)
- transnational crime
- human security and health security
- social, political and psychological aspects of security
- recognition and identification (e.g., optical imaging, biometrics, authentication and verification)
- smart surveillance systems
- applications of theoretical frameworks and methodologies (e.g., grounded theory, complexity, network sciences, modelling and simulation)

Together, the high-quality contributions to this series provide a cross-disciplinary overview of forefront research endeavours aiming to make the world a safer place.

The editors encourage prospective authors to correspond with them in advance of submitting a manuscript. Submission of manuscripts should be made to the Editor-in-Chief or one of the Editors.

More information about this series at http://www.springer.com/series/5540

Thirimachos Bourlai • Panagiotis Karampelas
Vishal M. Patel
Editors

Securing Social Identity in Mobile Platforms

Technologies for Security, Privacy and Identity Management

Springer

Editors
Thirimachos Bourlai
Lane Department of Computer Science
and Electrical Engineering
Multispectral Imagery Lab (MILab)
West Virginia University
Morgantown, WV, USA

Panagiotis Karampelas
Department of Informatics and Computers
Hellenic Air Force Academy
Acharnes Attica, Greece

Vishal M. Patel
Department of Electrical Engineering
Johns Hopkins University
Baltimore, MD, USA

ISSN 1613-5113 ISSN 2363-9466 (electronic)
Advanced Sciences and Technologies for Security Applications
ISBN 978-3-030-39488-2 ISBN 978-3-030-39489-9 (eBook)
https://doi.org/10.1007/978-3-030-39489-9

This Springer imprint is published by the registered company Springer Nature Switzerland AG
The registered company address is: Gewerbestrasse 11, 6330 Cham, Switzerland

Preface

This book presents novel research in the areas of social identity and security when using mobile platforms. The topics cover a broad range of applications related to securing social identity as well as latest advances in the field, including the presentation of novel research methods that are in the service of all citizens using mobile devices. More specifically, academic, industry-related, and government (law enforcement, intelligence, and defense) organizations will benefit from the research topics of this book that cover the concept of identity management and security using mobile platforms from various perspectives, i.e., whether a user navigates to social media, accesses their own phone devices and their bank accounts, uses online shopping service providers, accesses their personal documents or accounts with valuable information, surfs the Internet, or even becomes a victim of cyberattacks. In all of the aforementioned cases, there is a need for mobile-related technologies that protect users' social identity and well-being in the digital world, including the use of biometrics, cybersecurity software and tools, active authentication and identity anti-spoofing algorithms, and more.

The **first part** of the book covers a set of **mobile-based privacy and security technologies**, including the following topics, namely, "Shared Images and Camera Fingerprinting May Lead to Privacy Issues," "Adversarial Attacks in Mobile Environments," and "Personalized Data Minimization Assurance Using Bluetooth Low Energy."

The **second part** of the book covers a set of **biometrics technologies** including the following topics, namely, "On Designing a Forensic Toolkit for Rapid Detection of Factors that Impact," "Face Recognition Performance when Processing Large-Scale Face Datasets," "Classification of Soft Biometric Traits when Matching Near-Infrared Long-Range Face Images Against Their Visible Counterparts," "Quality and Match Performance Analysis of Band-Filtered Visible RGB Images," "Unconstrained Face Recognition Using Cell Phone Devices: Faces in the Wild," and "Face Detection in MWIR Spectrum."

The **third part** of the book covers a set of **mobile-based active authentication technologies** including the following topics, namely, "Mobile Active Authentication Based on Multiple Biometric and Behavioral Patterns," "Quickest Multiple User

Active Authentication," "Iris Recognition on Mobile: Real-Time Feature Extraction and Matching in the Wild," "A Protocol for Decentralized Biometric-Based Self-Sovereign Identity Ecosystem," and "Toward Wider Adoption of Continuous Authentication on Mobile Devices."

We hope this book can become a reference work for anyone working in the government, industry, or academia that uses technologies for security, privacy, and identity management. This book can also be used by researchers (academic or not) and master's and Ph.D. students who want to focus and be updated with the current developments on this area of research. Finally, we would like to thank all the contributors of the book for the high-quality work they have submitted to us and their support in the coordination of this publication.

Morgantown, WV, USA — Thirimachos Bourlai
Attica, Greece — Panagiotis Karampelas
Baltimore, MD, USA — Vishal M. Patel

Acknowledgments

In academic publications, assessing and selecting the best papers are always a challenging task for editors which is largely based on the voluntary work of academic experts in the research area of the publication. Finding these experts is not always easy due to their workload. Those who devoted their precious time to provide us with their constructive feedback and criticism deserve our undivided gratitude.

As the editors of the book, we would like to thank:

- Natalia Schmid, Suha Reddy Mokalla, and Jeremy Dawson from West Virginia University, USA
- Alexey Fartukov and Gleb Odinokikh from Samsung R&D Institute Rus, Russia
- Muhammad Nihal Hussain from the University of Arkansas at Little Rock, USA
- Asem Othman from Veridium, USA
- Ayman Abaza from USPTO, USA
- Kamer Vishi from the University of Oslo, Norway

Finally, we would also like to thank Annelies Kersbergen, Associate Editor, Security Science at Springer, for her continuous support, valuable feedback, and suggestions at all stages of the preparation of the book.

Contents

About the Editors and Contributors

Editors

Thirimachos Bourlai is an Associate Professor in the Lane Department of Computer Science and Electrical Engineering at WVU. He also serves as an Adjunct Assistant Professor in the WVU School of Medicine, Department of Ophthalmology, Department of Forensic and Investigative Science, and Department of Chemical Engineering (Biomedical Engineering). He is the Founder (2010) and Director of the Multispectral Imagery Lab.

After earning his Ph.D. in Face Recognition and completing a postdoctoral appointment at the University of Surrey (UK), he completed a second postdoc in a joint project between Methodist Hospital and the University of Houston in the fields of thermal imaging and human-based computational physiology. He joined the staff at WVU in 2009 serving as a Visiting Professor and later as a Research Assistant Professor in the Lane Department until August 2017.

Bourlai served and has been invited to serve as Chair at a number of biometrics conferences including ICB, BTAS, IJCB, FG, SPIE, ISS World Americas, IDGA, ACI, and the Biometrics Institute. He has served as a member of technical program committees for other primary computer vision- and biometrics-focused conferences. Several governmental agencies, organizations, and academic institutions have invited Bourlai to present his work, including the CIA, NSA, US Secret Service, US Army (various divisions), FBI, Amazon, SRC, Biometrics Institute, DSA, NLETS, IDGA, FedID, ConnectID, the Biometrics Summit Conference, the IEEE Signal Processing Society, the University of Notre Dame, the University of Pittsburgh, the University of Rochester, Rutgers University, the University of Houston, and the University of Newcastle (UK).

He is also a Series Editor of the Advanced Sciences and Technologies for Security Applications (book series by Springer, 2015–current); an Associate Editor of the IET journal *Electronics Letters*; an Editorial Manager of the *SPIE Newsroom Magazine*; an IEEE Biometrics Council VP on Education as of January 1, 2020; an Advisory Board Member (and Chair) of the IDGA Biometrics and Law

Enforcement Conference (yearly conference at Washington, DC); and a member of the Academic Research and Innovation Expert Group of the Biometrics Institute.

Panagiotis Karampelas completed his Ph.D. in Electronic Engineering from the University of Kent at Canterbury, UK, and his Master of Science degree from the Department of Informatics, National and Kapodistrian University of Athens with specialization in "High Performance Algorithms." He also completed his bachelor's degree in Mathematics from the same University majoring in Applied Mathematics. He has worked as an Associate Researcher in numerous European-funded research projects in the area of information society technologies program practicing his expertise in user interface design, usability evaluation, and application development. He has also worked for several years as a user interface designer, usability expert and senior developer in several IT companies, designing and implementing large-scale research and commercial information systems. He then joined Hellenic American University as Assistant Professor teaching human computer interaction, programming, managing information systems, and database management in the graduate and undergraduate programs. Currently, he is in the Department of Informatics and Computers, Hellenic Air Force Academy, teaching courses to pilots and engineers. His areas of interest include data mining, social network analysis, counterterrorism informatics, programming, mobile application development, information visualization, human-computer interaction, artificial neural networks, power transmission, and distribution systems. He has published a number of books and research articles in his major areas of interests in international journals and conferences. He serves as a Series Editor in the Book Series Advanced Sciences and Technologies for Security Applications and as an Associate Editor in the journal *Social Network Analysis and Mining*. He also serves as program committee member in a large number of scientific journals and international conferences in his fields of interests such as the European Intelligence and Security Informatics Conference (EISIC) and the IEEE/ACM International Conference on Advances in Social Networks Analysis and Mining (ASONAM).

Vishal M. Patel is an Assistant Professor in the Department of Electrical and Computer Engineering (ECE) at Johns Hopkins University. Prior to joining Hopkins, he was an A. Walter Tyson Assistant Professor in the Department of ECE at Rutgers University and a member of the research faculty at the University of Maryland Institute for Advanced Computer Studies (UMIACS). His current research interests include signal processing, computer vision, and pattern recognition with applications in biometrics and imaging. He has received a number of awards including the 2016 ONR Young Investigator Award, the 2016 Jimmy Lin Award for Invention, A. Walter Tyson Assistant Professorship Award, Best Paper Award at IEEE AVSS 2017 and 2019, Best Paper Award at IEEE BTAS 2015, Honorable Mention Paper Award at IAPR ICB 2018, two Best Student Paper Awards at IAPR ICPR 2018, and Best Poster Awards at BTAS 2015 and 2016. He is an Associate Editor of the IEEE *Signal Processing Magazine*, IEEE Biometrics Compendium, and *Pattern Recognition* journal and serves on the Information Forensics and Security Technical Committee of the IEEE Signal Processing Society. He serves as the Vice President

(Conferences) of the IEEE Biometrics Council and is a member of Eta Kappa Nu, Pi Mu Epsilon, and Phi Beta Kappa.

Contributors

Alejandro Acien Biometrics and Data Pattern Analytics (BiDA) Lab, EPS, Universidad Autonoma de Madrid, Madrid, Spain

Flavio Bertini Department of Computer Science and Engineering, University of Bologna, Bologna, Italy

Thirimachos Bourlai Lane Department of Computer Science and Electrical Engineering, Multispectral Imagery Lab (MILab), West Virginia University, Morgantown, WV, USA

John Callahan Veridium IP Ltd, Boston, MA, USA

Jason Crakes Lane Department of Computer Science and Electrical Engineering, West Virginia University, Morgantown, WV, USA

Jeremy Dawson Lane Department of Computer Science and Electrical Engineering, West Virginia University, Morgantown, WV, USA

Alexey Fartukov Samsung R&D Institute Russia, Moscow, Russia

Julian Fierrez School of Engineering, Universidad Autonoma de Madrid, Madrid, Spain

John Goodwyn Lane Department of Computer Science and Electrical Engineering, West Virginia University, Morgantown, WV, USA

Zafeiria-Marina Ioannou Computer Engineering and Informatics, Department, University of Patras, Patras, Greece

Michael Martin West Virginia University, Morgantown, WV, USA

S. Means Lane Department of Computer Science and Electrical Engineering, West Virginia University, Morgantown, WV, USA

Suha Reddy Mokalla West Virginia University, Morgantown, WV, USA

Danilo Montesi Department of Computer Science and Engineering, University of Bologna, Bologna, Italy

Aythami Morales Biometrics and Data Pattern Analytics (BiDA) Lab, EPS, Universidad Autonoma de Madrid, Madrid, Spain

Neeru Narang Lane Department of Computer Science and Electrical Engineering, West Virginia University, Morgantown, WV, USA

Tempestt Neal Department of Computer Science and Engineering, University of South Florida, Tampa, FL, USA

Gleb Odinokikh Samsung R&D Institute Russia, Moscow, Russia

Asem Othman Veridium IP Ltd, Boston, MA, USA

Vishal M. Patel Department of Electrical Engineering, Johns Hopkins University, Baltimore, MD, USA

Pramuditha Perera Department of Electrical and Computer Engineering, Johns Hopkins University, Baltimore, MD, USA

Sanka Rasnayaka School of Computing, National University of Singapore, Singapore, Singapore

Jacob Rose West Virginia University, Morgantown, WV, USA

Rahimeh Rouhi Department of Computer Science and Engineering, University of Bologna, Bologna, Italy

Evangelos Sakkopoulos Department of Informatics, University of Piraeus, Piraeus, Greece

Terence Sim School of Computing, National University of Singapore, Singapore, Singapore

Ruben Vera-Rodriguez Biometrics and Data Pattern Analytics (BiDA) Lab, EPS, Universidad Autonoma de Madrid, Madrid, Spain

Emmanouil Viennas Computer Engineering and Informatics, Department, University of Patras, Patras, Greece

Damon Woodard Department of Electrical and Computer Engineering, University of Florida, Gainesville, FL, USA

Part I
Mobile-Based Privacy & Security

Shared Images and Camera Fingerprinting May Lead to Privacy Issues

Rahimeh Rouhi, Flavio Bertinil, and Danilo Montesi

Abstract Social networks have become an indispensable part of everyday life by providing users with different types of interaction. However, sharing different types of data, such as text, image, video and etc., on social networks, gives rise to user privacy concerns and risks, while the user is not aware of that. In this chapter, we show how the images shared by users can be applied to fingerprint the acquisition devices and link user profiles on social networks.

Keywords Camera fingerprinting · Social networks · User profile linking · User privacy

1 Introduction

In recent years, social networks have revolutionized the web by providing users with easy and inexpensive types of interactions [1], e.g., by sending texts and sharing images and videos. Many social networks have introduced applications particularly dedicated to major mobile devices (e.g., smartphones), introducing changes in user habits regarding multimedia content on social networks [2]. The ever increasing use of smartphones has led users to take more and more digital images and share them across various social networks [3]. On the other hand, the users may not be aware of putting their privacy at risk. Privacy issues make social networks to protect users' privacy by anonymizing the shared data by the users. For example, some social networks remove the metadata information from the file header of images and the information is not even available from Exchangeable Image File Format (EXIF) [4]. *Is this solution sufficient to protect users' privacy?*

In this chapter, we show how the images shared by users can be applied to fingerprint the acquisition devices and link user profiles on social networks,

R. Rouhi (✉) · F. Bertini · D. Montesi
Department of Computer Science and Engineering, University of Bologna, Bologna, Italy
e-mail: rahimeh.rouhi2@unibo.it; flavio.bertini2@unibo.it; danilo.montesi@unibo.it

T. Bourlai et al. (eds.), *Securing Social Identity in Mobile Platforms*, Advanced Sciences and Technologies for Security Applications,
https://doi.org/10.1007/978-3-030-39489-9_1

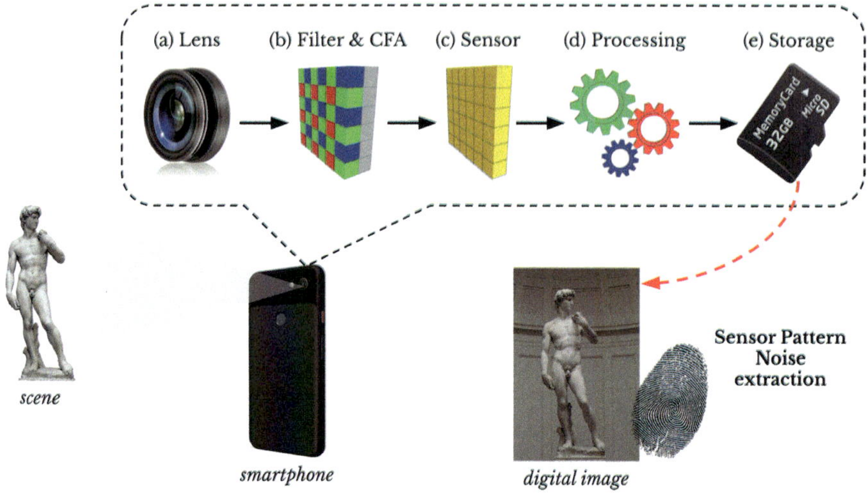

Fig. 1 Different components embedded in camera

which may lead to privacy issues. Many approaches have been proposed to get smartphone fingerprints using a variety of built-in sensors such as accelerometers [5], gyroscopes [6], magnetometers [7, 8], cameras [9], and paired microphones and speakers [10]. All of them have hardware imperfections, which are created during the manufacturing process, and can be used to fingerprint each device. As it can been seen from Fig. 1, digital cameras have mainly several built-in components such as (a) lens, (b) Color Filter Arrays (CFA) and (c) sensors. More specifically, lens produces a similar-prism phenomenon and divides the light beam into a spectrum of rainbow colors. This causes a shift in the point where different wavelengths (colors) converge, that is the characteristic of the lens. Optical anti-aliasing filter and CFA are in front of the image sensor and reconstruct the color information. The color subsampling is affected by noise [11]. The camera fingerprint formed by sensor imperfections has been known as a reliable characteristic making a smartphone trackable [12–14]. Particularly, the Sensor Pattern Noise (SPN), due to camera sensor imperfections is considered as a unique characteristic to fingerprint a source camera [12].

The SPN contains the Fixed Pattern Noise (FPN) and the Photo-Response Non-Uniformity (PRNU) noise, see Fig. 2. The FPN, which is created by dark currents, is the pixel-to-pixel differences when the sensor array is not exposed to light. Since the FPN is an additive noise, some cameras suppress it automatically, by subtracting a dark frame from every captured image. The FPN is affected by ambient temperature and exposure. The dominant component in the SPN is the PRNU noise. It is generated primarily by Pixel Non-Uniformity (PNU) defined as different sensitivity of pixels to light, which is caused by the inhomogenity of silicon wafers and imperfections. The character and origin of the PNU noise make it unlikely that even sensors from the same wafer would present correlated PNU patterns. So,

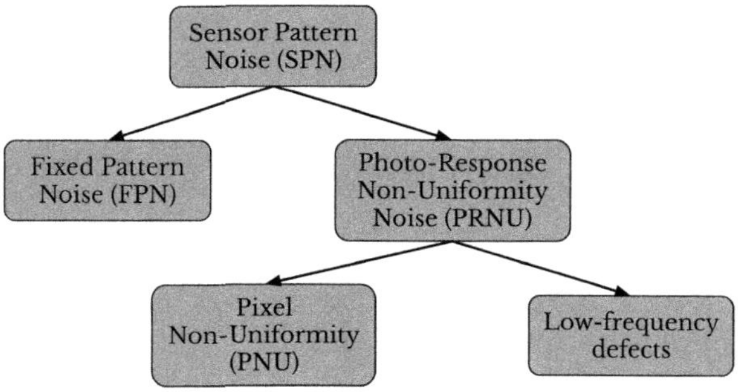

Fig. 2 Pattern noise of camera sensor

the PNU noise is not dependent on temperature or humidity. Light refraction on dust particles and optical surfaces and zoom settings also contribute to the PRNU noise. These components are of low spatial frequency in nature and they are not a characteristic of the sensor. Hence, only the PNU component, which is an intrinsic characteristic of the sensor, is used for fingerprinting sensors [15]. The PRNU is a strong tool for fingerprinting smartphones as it is unique for an individual device. Besides, it is stable against environmental conditions [16].

Given a set of images captured by a specific smartphone, the SPN can be approximated by averaging the residual noises existing in the images [12]. The residual noise is the difference between the image content and its denoised version obtained by a denoising filter. To fingerprint the acquisition camera sources of the shared images, smartphone identification is defined, which deals with 1-to-m matching problem and determines which smartphone out of m took a given image [17]. In smartphone identification, the number of the acquisition smartphones has to be known. However, when the number of the camera sources generated the images is unknown, a SPN-based image clustering is needed [18]. The aim of clustering is to group the residual noises into several clusters, each of them includes the residual noises sharing the same SPN characteristics. Once the cameras are fingerprinted, with having the profile tags, specifying each user on a social networks platform, the profiles sharing images from the same source are linked. This is called User Profile Linking (UPL) [19, 20]. It can be achieved within the same social network, i.e., intra-layer UPL [21] or across social networks, i.e., inter-layer UPL [22]. In this chapter, some proposed methods in the literature for smartphone identification, SPN-based image clustering and UPL are mentioned, and their results are reported.

2 Background

2.1 Pre-processing

First of all, since images come from different devices, some pre-processing is applied to images. As the SPN is dependent on the orientation of images, the correct

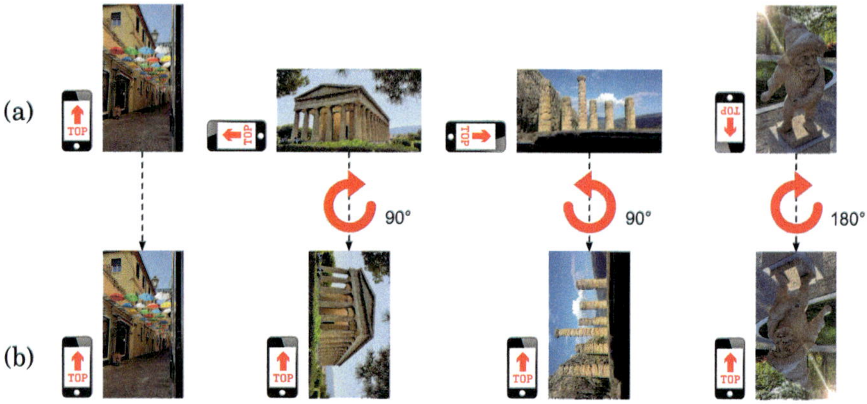

Fig. 3 Normalization of images' orientation: (**a**) the smartphones' orientation along with the resulting images, and (**b**) the corresponding images after rotation

image orientation has to be provided. Once an image is saved, its orientation may be changed. By the EXIF tool, the metadata information of images is obtained to get their correct orientation. Then, all the images can be rotated to either portrait or landscape orientation [23], see Fig. 3. Some smartphone setting and social network platforms remove the orientation information from the file header of images, due to the user privacy. However, for those images whose orientation information is lost there are still some ways to tackle the orientation issue. For example, in [21], once the similarity between residual noises is calculated, one of the residual noises is rotated 180° and again the similarity is computed. Next, the highest value is considered as the similarity between the residual noises. This approach is time consuming and is not applicable to large-scale datasets. However, in [24], a rotation-invariant binary representation of SPN was proposed, which reduces the computational cost.

Digital images can be represented in different color spaces, commonly RGB and YCbCr. A color RGB image consists of three channels namely Red (R), Green (G), and Blue (B), while YCbCr represents color images as brightness/luma (Y) and two color difference signals, i.e., blue minus luma (Cb) and red minus luma (Cr). The luma component in YCbCr color space is essentially the grayscale copy of the image. Smartphones are equipped with RGB camera, however, the YCbCr representation can be obtained through a mathematical coordinate transformation from the associated RGB color space. The SPN extraction can be performed using all these different channels. To reduce memory usage and the computational cost, a specific channel should be considered. Based on our previous work [21], among the components, the Y channel resulted in the best effectiveness in camera fingerprinting.

Dark and saturated images do not provide trustworthy SPN [25]. Considering these images makes the clustering task unreliable and computationally expensive [18]. Hence, these images are removed. The value of each pixel in a grayscale image can be in the range [0–255], where the values 0 and 255 represent black and white pixel intensity values, respectively. Accordingly, the image whose 70% of pixel intensities are smaller than 50 or greater than 250 is considered as a dark or saturated image, respectively.

2.2 *Sensor Pattern Noise Extraction*

Each residual noise is the difference between the image content and its denoised version acquired by a denoising filter. The residual noises are extracted from the pre-processed images as follows:

$$RN = I - d(I) \quad (1)$$

where I and $d()$ are an image and a denoising filter, respectively. Then, by averaging the residual noises extracted from n images taken by a given smartphone, the SPN, i.e., the camera fingerprint, can be approximated by:

$$SPN = \frac{1}{n}\sum_{j=1}^{n} RN_j \quad (2)$$

The quality of the extracted residual noise and SPN is dependent on $d()$ and n. As the denoising filter, Block-Matching and 3D (BM3D) introduced by [26], results in better quality of SPNs. Through BM3D, non-unique artifacts are removed by using zero-meaning all columns and rows, and Wiener filtering in the Fourier domain [25].

The process of content compression performed by social networks causes loss of image details and weakens the SPN, so it is desirable to apply high resolutions of residual noises. The heavy overhead on data storage and computation limits using the residual noises with high resolutions, especially for clustering large-scale datasets. In our works [21, 22, 27], after extracting the residual noises from the full-size grayscale images, we resize them to a specific resolution by bicubic interpolation [28], unlike many works which crop the central block of residual noises. By resizing, the lowest and the highest resolutions can be upscaled and downscaled, respectively, to a specific size for more efficient use of available information, see Fig. 4. Resizing is a flexible way to calculate the similarities between the residual noises with different resolutions. Given images with different resolutions such as 2560 × 1920 and 960 × 720 px, to calculate similarities, all the extracted residual noises from the images are typically cropped to the lowest resolution, i.e., 960 × 720 px in this case. Hence, a large part of residual noises with the highest resolution, i.e., 2560 × 1920 px, is discarded. In contrast, by resizing, more efficient use of available information is provided. Although zero-padding can be another option to handle the computation of similarities between residual noises with different resolutions, it may have its own issues, e.g., memory usage.

Fig. 4 (**a**) Cropping versus (**b**) resizing

2.3 Classification and Clustering

Classification and clustering are the two types of learning methods which organize objects into groups based on one or more features. They appear to be similar, but they are different in the context of data mining. Classification is the process of learning a model that elucidate different predetermined classes of data. It is a two-step process composed of a learning step and a classification step. In the learning, a classification model is constructed, and it is trained in a supervised approach, such that predefined labels are assigned to objects by features. Then, the trained classifier is given the objects whose labels are unknown, to assign them a label as their class.

On the contrary, clustering is performed in an unsupervised learning approach where similar objects are grouped, based on their features. It does not involve training or learning, and the training sample is not known previously. It organizes objects into clusters where the objects reside inside a cluster will have high similarity and the objects of two clusters would be dissimilar to each other. Here the two clusters can be considered as disjoint [29, 30].

2.4 Similarity Measure for Camera Fingerprints

The classification and clustering are performed on the similarities between the camera fingerprints, whether residual noises or SPNs. The Normalized Cross Correlation (NCC) similarity between any two camera fingerprints $f_i = [x_1, \ldots, x_l]$ and $f_j = [y_1, \ldots, y_l]$ is calculated as follows:

$$\mathcal{A}(f_i, f_j) = \frac{\sum_{n=1}^{l}(x_n - \overline{f}_i)(y_n - \overline{f}_j)}{\sqrt{\sum_{n=1}^{l}(x_n - \overline{f}_i)^2 \sum_{n=1}^{l}(y_n - \overline{f}_j)^2}} \tag{3}$$

where $\overline{f}_i$ and $\overline{f}_j$ represent the means of the two fingerprints, respectively.

Because of the varying qualities of SPNs of different cameras, the average correlation between the residual noises from one camera may differ from that of other camera [31]. This makes the clustering of SPNs more challenging. To address this issue, an alternative similarity measure is calculated based on Shared κ-Nearest Neighbors (SNN) proposed by [32]:

$$\mathcal{W}(a_i, a_j) = |\mathbb{N}(\rho_i) \cap \mathbb{N}(\rho_j)| \tag{4}$$

where ρ_i and ρ_j are two values in the correlation matrix $\mathcal{A}$ in (3), and $\mathbb{N}(\rho_i)$ and $\mathbb{N}(\rho_j)$ are the SNN of ρ_i and ρ_j, so $\mathcal{W}(a_i, a_j)$ results in the number of κ-nearest neighbours shared by ρ_i and ρ_j. In other words, each element in the correlation matrix, i.e., $\mathcal{A}(a_i, a_j)$, is mapped to $\mathcal{W}(a_i, a_j)$. Then, clustering is performed on the resulted matrix $\mathcal{W}$ from SNN.

3 Shared Image Analysis

The SPN caused by camera sensor imperfections remains stable as the residual noises in the images shared by users on their profiles. By analyzing the residual noises, the smartphone generated the images can be fingerprinted and the profiles sharing the images on social network platforms can be linked. That may give rise to user privacy issues. As it is shown in Fig. 5, even if there is not a friendship between

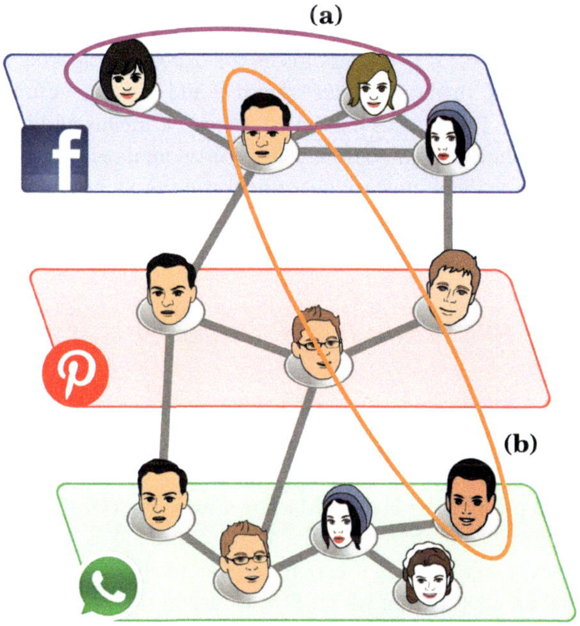

Fig. 5 Using shared images and fingerprinting the acquisition smartphones may lead to users' privacy issues by: (**a**) intra-layer UPL and (**b**) inter-layer UPL

users on social networks, their profiles can be linked by the shared images within the same or across different social networks, i.e., intra-layer UPL and inter-layer UPL, respectively.

3.1 Smartphone Identification

Smartphone identification is the task of identifying the source cameras generated the images, and it can be achieved by clustering. Let I be a set including n images shared by a user on a social network platform, and $S = \{S_1, S_2, \ldots, S_m\}$ be a set of m camera sources generated the images. The aim is to group the images of I into the right sources of S, where each camera source S_i has its own set of images, i.e., $I_{\langle 1,i \rangle}, \ldots, I_{\langle j,i \rangle}, \ldots, I_{\langle n,i \rangle} \in S_i$, where $I = \bigcup I_{\langle i,j \rangle}$, $\forall\, i = 1, \ldots, n$ and $j = 1, \ldots, m$.

The aim of clustering is to group the residual noises, extracted from the images, into some clusters such that the residual noises in the same cluster have more similarity compared with those in different clusters. The traditional clustering techniques such as hierarchical [33], k-means [34] and k-medoids [35] have been frequently applied to different fields of science. Hierarchical clustering typically organizes the objects into a dendrogram. The dendrogram is a tree structure whose leaf nodes, middle nodes and root represent, respectively, the individual data, merged groups of similar objects, and all objects together [33]. In partitional clustering, i.e., k-means and k-medoids, the objects are divided into some partitions, each of which is considered as a cluster. The partitional clustering starts by initializing a set of k cluster centers. Then, each object is assigned to the cluster whose center is the nearest [36, 37]. K-medoids is an expensive approach, but it is a more reliable technique in the presence of noise and outliers compared to the other clustering methods [38]. In [21], the hierarchical, k-means and k-medoids have been used for smartphone identification. The experimental results have shown the effectiveness of the k-medoids clustering in fingerprinting smartphones, even for the images from identical models of smartphones.

3.2 SPN-Based Image Clustering

The traditional clustering algorithms have to be provided by the initial information about the number of camera sources. In the cases without the initial information, usually combining different clustering algorithms is more effective. In [27], we have presented a Hybrid Markov Clustering (HMC) algorithm to group the images captured and shared by the users on social network platforms. Particularly, the HMC exploits image resizing, hierarchical and graph-based clustering algorithms, and an adaptive threshold [18] to group the images. We have shown that resizing the residual noises to a specific resolution results in better qualities of the sensor pattern noises and better effectiveness for clustering. By using Markov clustering, the hierarchical clustering is conducted in such a way that the representative clusters with a higher probability of belonging to the same camera are selected for merging. The adaptive threshold for merging the representative clusters depends on the quality of the obtained clusters. The threshold that is updated during the hierarchical approach can produce more precise clusters even for images from the same model of smartphones. The HMC is applicable to large-scale datasets because it partitions the dataset into batches to address the problem of memory constraint for loading many residual noises into RAM. It exploits the inherent sparseness of correlation matrix [18], and checks only the candidate clusters for a merging. The candidate clusters are introduced by the Markov Clustering algorithm and a nearest neighboring. This accelerates the clustering by calculating only a small portion (about 15%) of the full-pairwise correlation matrix. The experimental results confirm the effectiveness and efficiency of the proposed algorithm in comparison with the state-of-the-art algorithms.

3.3 User Profile Linking

After clustering the residual noises extracted from the shared images on a social network platform, with having the profile tags of users, the profiles sharing images from the same source are linked. Especially, it results in intra-layer UPL shown in Fig. 5a. In [22], an inter-layer UPL approach was proposed, where the clustering is followed by a classification stage based on Artificial Neural Network (ANN), to match the residual noises from one social network with the fingerprints obtained from the clustering. More specifically, the correlation values between the residual noises and the SPNs are computed and a similarity matrix is formed. The matrix is used for training and test the ANN through a 10-folds cross validation evaluation [39]. In the test, the trained ANN is provided by 10% of the rows in the correlation matrix to classify each image. By using 10-fold cross validation, all the samples in the correlation matrix are tested as there is a swap between training and test in each iteration. With the results of classification and the profile tags of users, the profiles sharing images from the same source camera, on different social networks are linked, called inter-layer UPL shown in Fig. 5b.

3.4 Experimental Results

In the implementations, VISION image dataset introduced by [23] is applied, which includes 7480 Native (N) images from 35 identical and also different smartphone models and brands. The images were shared on social networks such as WhatsApp (W), Facebook High (FH) resolution and Facebook Low (FL) resolution. They are depicted by $\mathcal{D}^{\mathrm{N}}$, $\mathcal{D}^{\mathrm{W}}$, $\mathcal{D}^{\mathrm{FH}}$ and $\mathcal{D}^{\mathrm{FL}}$, respectively.

The clustering and classification can be validated by different measures. Four definitions based on the agreement between two sets of labels, i.e., ground truth or target clusters $T = \{t_1, t_2, \ldots, t_g\}$ and the resulted clusters $C = \{c_1, c_2, \ldots, c_o\}$ are available, where g and o are the number of the target and the resulted clusters, respectively. Given two samples d_i and a_j in a dataset $\mathcal{D} = \{a_1, a_2, \ldots, a_N\}$, we have the following definitions:

i. True Positive: $TP = \{(a_i, a_j) : t_i = t_j \ and \ c_i = c_j\}$, which means that the two samples a_i and a_j are from the same cluster in T, and they are also in the same output cluster in C.
ii. False Negative: $FN = \{(a_i, a_j) : t_i = t_j \ and \ c_i \neq c_j\}$, which means that the two samples a_i and a_j are from the same cluster in T, while they are not in the same cluster in C.
iii. False Positive: $FP = \{(a_i, a_j) : t_i \neq t_j \ and \ c_i = c_j\}$, which means that the two samples a_i and a_j are not from the same cluster in T, but they are in the same output cluster in C.
iv. True Negative: $TN = \{(a_i, a_j) : t_i \neq t_j \ and \ c_i \neq c_j\}$, which means that the two samples a_i and a_j are not from the same cluster in T, and they are also not in the same cluster in C.

Regarding the above definitions, *Precision rate* $\mathcal{P}$, *Recall rate* $\mathcal{R}$ also known as *True Positive Rate* (*TPR*), *F1-measure* $\mathcal{F}$, *Rand Index* (*RI*), *Adjusted Rand Index* (*ARI*), *Purity* and *False Positive Rate* (*FPR*) are depicted by (5)–(11):

$$\mathcal{P} = \frac{|TP|}{|TP| + |FP|} \tag{5}$$

$$\mathcal{R} = \frac{|TP|}{|TP| + |FN|} \tag{6}$$

$$\mathcal{F} = 2 \cdot \frac{\mathcal{P} \cdot \mathcal{R}}{\mathcal{P} + \mathcal{R}} \tag{7}$$

$$RI = \frac{|TP| + |TN|}{|TP| + |FP| + |TN| + |FN|} \tag{8}$$

where $|.|$ shows the number of the pairs in the corresponding set defined in (i)–(iv). The value of *RI* varies between 0 and 1, respectively showing no agreement and full agreement between the clustering results and the ground truth. For two random clusters, the average of $\overline{RI}$ is a non-zero value. To get rid of this bias, *ARI* was proposed by [40]:

$$ARI = \frac{RI - \overline{RI}}{1 - \overline{RI}} \tag{9}$$

Purity and *FPR* are defined as follows:

$$Purity = \frac{\sum_{i=1}^{|C|} \frac{|\widehat{c_i}|}{|c_i|}}{|C|} \tag{10}$$

where $|C|$ is the number of the obtained clusters, $\widehat{c_i}$ denotes the number of residual noises with the dominant cluster label in the cluster c_i, and $|c_i|$ is the total number of residual noises in c_i.

$$FPR = \frac{|FP|}{|FP + TN|} \tag{11}$$

Also, for clustering evaluation, the ratio of the number of the obtained clusters n_o over the number of ground truth clusters n_g is computed:

$$\mathcal{N} = \frac{n_o}{n_g} \tag{12}$$

The results of resizing versus cropping the residual noises are presented in Tables 1 and 2, respectively, for *original-by-original SI* and HMC methods. The best value of each measure is highlighted in bold. The experiments were done on the sample dataset $\mathcal{D}_0^N \subseteq \mathcal{D}^N$. From each of 35 smartphones in $\mathcal{D}^N$, 100 images are randomly selected, so $\mathcal{D}_0^N$ includes 3500 images. The sample dataset makes the resolution setting facilitative and it still includes images from a variety of smartphone models and brands. As it can be seen from the tables, the resolution 1024×1024 results in the best effectiveness among the other resolutions.

In Fig. 6, the correlation matrices for the dataset $\mathcal{D}^N$ are shown. The first matrix shown in (a) is the full-pairwise correlation matrix, while the ones in (c) and (e) are respectively the result of the mapping the matrix in (a) by SNN and the computed correlation matrix by HMC. In sub-figures (b), (d) and (f), the probability distributions corresponding to the matrices can be seen. The full-pairwise correlation includes intra-camera and inter-camera correlation values. While the intra-camera correlations are related to the similarities between residual noises

Table 1 Results (%) of resizing versus cropping residual noises in *original-by-original SI* on $\mathcal{D}_0^N$, by testing different resolutions

	Resizing						Cropping[a]					
Size	$\mathcal{P}$	$\mathcal{R}$	$\mathcal{F}$	*ARI*	*Purity*	*FPR*	$\mathcal{P}$	$\mathcal{R}$	$\mathcal{F}$	*ARI*	*Purity*	*FPR*
1280 × 1024	0.84	0.85	0.85	**0.86**	**0.92**	**0.00**	—	—	—	—	—	—
1024 × 1024	**0.85**	**0.87**	**0.86**	**0.86**	**0.92**	**0.00**	—	—	—	—	—	—
960 × 720	0.84	**0.87**	**0.86**	0.85	0.91	**0.00**	0.84	0.86	0.85	0.85	0.91	**0.00**
512 × 512	0.78	0.81	0.79	0.79	0.87	**0.00**	0.81	0.83	0.82	0.81	0.89	**0.00**
256 × 256	0.19	0.22	0.20	0.18	0.46	0.02	0.63	0.65	0.64	0.63	0.77	0.01
128 × 128	0.03	0.04	0.03	0.00	0.93	0.03	0.32	0.33	0.33	0.31	0.54	0.02

[a]The highest resolution for cropping is 960 × 720 px corresponding to the highest image resolution in $\mathcal{D}_0^N$

Table 2 Results (%) of resizing versus cropping residual noises in HMC on $\mathcal{D}_0^N$, by testing different resolutions

	Resizing							Cropping[a]						
Size	$\mathcal{P}$	$\mathcal{R}$	$\mathcal{F}$	*ARI*	*Purity*	*FPR*	$\mathcal{N}$	$\mathcal{P}$	$\mathcal{R}$	$\mathcal{F}$	*ARI*	*Purity*	*FPR*	$\mathcal{N}$
1280 × 1024	**0.99**	0.75	**0.86**	0.85	**0.99**	**0.00**	**35/35**	—	—	—	—	—	—	– /35
1024 × 1024	**0.99**	**0.76**	**0.86**	**0.86**	**0.99**	**0.00**	**35/35**	—	—	—	—	—	—	– /35
960 × 720	0.98	0.72	0.83	0.83	**0.99**	**0.00**	34/35	0.69	0.61	0.65	0.64	0.96	**0.00**	31/35
512 × 512	0.95	0.44	0.60	0.59	0.96	**0.00**	48/35	0.67	0.49	0.57	0.56	0.96	**0.00**	37/35
256 × 256	0.50	0.02	0.05	0.04	0.59	**0.00**	280/35	0.65	0.30	0.41	0.40	0.88	**0.00**	60/35
128 × 128	0.031	0.14	0.05	0.00	0.17	0.13	26/35	0.48	0.13	0.21	0.20	0.59	**0.00**	159/35

[a]The highest resolution for cropping is 960 × 720 px corresponding to the highest image resolutions in $\mathcal{D}_0^N$

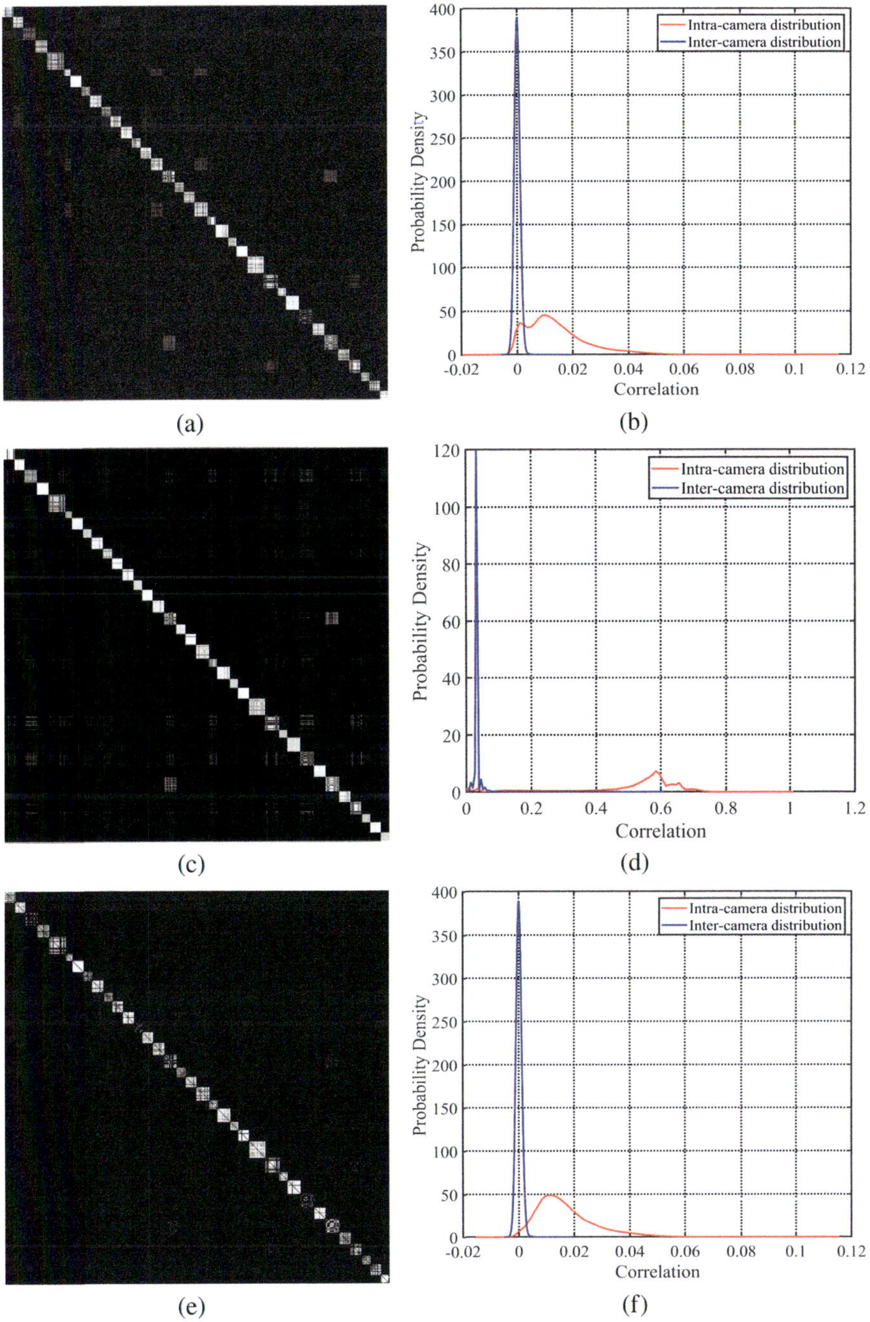

Fig. 6 Pairwise similarities of residual noises: (**a**) correlation values (**c**) mapped correlation values by SNN, and (**e**) computed correlation values by HMC, and (**b**), (**d**) and (**f**) are their corresponding probability distributions

from the same smartphones, the inter-camera correlations are the ones between residual noises from different smartphones. They are represented in the full-pairwise correlation matrix in sub-figure (a) as the diagonal and off-diagonal parts, respectively. Generally, the more the intra-camera and inter-camera correlations are separable, the better the quality of the clustering is obtained. Accordingly, comparing the sub-figs (f) and (d) with the one in (b) shows the impact of SNN and MCL on the computed correlation. The results of the methods on different datasets are reported in Tables 3 and 4. The best and the worst results are related to $\mathcal{D}^{N}$ and $\mathcal{D}^{FL}$, respectively, since the resolutions of the native images is higher than those of FL. The results of the dataset $\mathcal{D}^{N}$ are also reported as a benchmark for the proposed methods because native images have the highest resolutions.

The results of inter-layer UPL for all the possible pairs of social networks are presented in Table 5. Using images in $\mathcal{D}^{W}$ to classify the images in the other datasets i.e., $\mathcal{D}^{FH}$ and $\mathcal{D}^{FL}$ concluded the best results. It is interesting that the classification

Table 3 Intra-layer UPL based on smartphone identification

Dataset	$\mathcal{P}$	$\mathcal{R}$	$\mathcal{F}$	*ARI*	*Purity*	*FPR*
$\mathcal{D}^{N}$	0.827	0.834	0.831	0.825	0.894	0.005
$\mathcal{D}^{W}$	0.742	0.751	0.746	0.738	0.839	0.007
$\mathcal{D}^{FH}$	0.700	0.729	0.714	0.705	0.793	0.009
$\mathcal{D}^{FL}$	0.412	0.424	0.418	0.400	0.573	0.018

Table 4 Intra-layer UPL based on hybrid clustering

Dataset	$\mathcal{P}$	$\mathcal{R}$	$\mathcal{F}$	*ARI*	*Purity*	*FPR*	$\mathcal{N}$
$\mathcal{D}^{N}$	0.992	0.720	0.834	0.830	0.994	0.000	37/35
$\mathcal{D}^{W}$	0.964	0.600	0.733	0.727	0.975	0.000	33/35
$\mathcal{D}^{FH}$	0.962	0.610	0.746	0.740	0.975	0.000	33/35
$\mathcal{D}^{FL}$	0.750	0.513	0.609	0.599	0.847	0.005	33/35

Table 5 Inter-layer UPL: $\mathcal{D}^{j} - \mathcal{D}^{i}$ denotes the images in $\mathcal{D}^{j}$ are classified based on the clustered images in $\mathcal{D}^{i}$

Dataset	$\mathcal{P}$	$\mathcal{R}$	$\mathcal{F}$	*ARI*	*Purity*	*FPR*
$\mathcal{D}^{W} - \mathcal{D}^{N}$	0.882	0.836	0.829	0.824	0.910	0.005
$\mathcal{D}^{FH} - \mathcal{D}^{N}$	0.795	0.821	0.808	0.802	0.900	0.006
$\mathcal{D}^{FL} - \mathcal{D}^{N}$	0.730	0.774	0.752	0.744	0.883	0.008
$\mathcal{D}^{N} - \mathcal{D}^{W}$	0.778	0.807	0.792	0.786	0.907	0.006
$\mathcal{D}^{FH} - \mathcal{D}^{W}$	0.755	0.785	0.772	0.762	0.878	0.007
$\mathcal{D}^{FL} - \mathcal{D}^{W}$	0.755	0.782	0.775	0.760	0.878	0.007
$\mathcal{D}^{N} - \mathcal{D}^{FH}$	0.756	0.798	0.776	0.769	0.900	0.007
$\mathcal{D}^{W} - \mathcal{D}^{FH}$	0.755	0.781	0.772	0.761	0.877	0.007
$\mathcal{D}^{FL} - \mathcal{D}^{FH}$	0.754	0.776	0.760	0.756	0.871	0.007
$\mathcal{D}^{N} - \mathcal{D}^{FL}$	0.632	0.665	0.648	0.638	0.772	0.011
$\mathcal{D}^{W} - \mathcal{D}^{FL}$	0.589	0.610	0.59	0.582	0.723	0.013
$\mathcal{D}^{FH} - \mathcal{D}^{FL}$	0.585	0.611	0.600	0.586	0.736	0.012

of the images in $\mathcal{D}^{FL}$ in inter-layer UPL compared with the clustering the images in intra-layer UPL, see Tables 3 and 4, generates better results in fingerprinting smartphones.

4 Concluding Remarks

It has been shown that even with removing the file header of images, there are still some information in the shared images by users on their profiles, and it can be extracted to fingerprint the acquisition cameras and even to do UPL. In particular, it is possible to link user profiles on different social networks without using any personal or privacy-sensitive data of the user. The quality of smartphone fingerprints depends on the resolution of the images. The less the image resolution is lost during the uploading process on social networks, the better the quality of residual noise and subsequently camera fingerprint and UPL are obtained.

5 Exercises

1. Download a set of images, native or shared images on one social network, taken by 18 smartphones available via the link http://smartdata.cs.unibo.it/datasets#images, and try to apply the following steps to the dataset:
   i. Extract residual noises from the images, by (1) and (2), the code of BM3D can be downloaded from http://www.cs.tut.fi/~foi/GCF-BM3D/.
   ii. Compute the similarity matrix of the extracted residual noises, by (3).
   iii. Use a clustering algorithm with known or unknown number of smartphones.
   iv. Validate the clustering results based on different measures, by (5)–(11).
2. How do the results change if filters, e.g., smartphone camera or Instagram filters, are applied to the images?
3. How is the camera fingerprinting affected if the captured images by a smartphone undergo some geometry transformations like cropping, resizing or rotation?

References

1. Tuunainen VK, Pitkänen O, Hovi M (2009) Users' awareness of privacy on online social networking sites-case facebook. In: Bled 2009 proceedings, p 42
2. Norouzizadeh Dezfouli F, Dehghantanha A, Eterovic-Soric B, Choo K-KR (2016) Investigating social networking applications on smartphones detecting facebook, twitter, linkedin and google+ artefacts on android and ios platforms. Aust J Forensic Sci 48(4):469–488
3. Liu Q, Li X, Chen L, Cho H, Cooper P, Chen Z, Qiao M, Sung A (2012) Identification of smartphone-image source and manipulation. In: Advanced research in applied artificial

intelligence, pp. 262–271
4. JEITA (2002) Exchangeable image file format for digital still cameras: Exif version 2.2. http://www.exif.org/Exif2-2.PDF
5. Dey S, Roy N, Xu W, Choudhury RR, Nelakuditi S (2014) Accelprint: imperfections of accelerometers make smartphones trackable. In: NDSS
6. Willers O, Huth C, Guajardo J, Seidel H (2016) Mems-based gyroscopes as physical unclonable functions. In: IACR Cryptology ePrint Archive, vol 2016, p 261
7. Jin R, Shi L, Zeng K, Pande A, Mohapatra P (2016) Magpairing: pairing smartphones in close proximity using magnetometers. IEEE Trans Inf Forensics Secur 11(6):1306–1320
8. Amerini I, Becarelli R, Caldelli R, Melani A, Niccolai M (2017) Smartphone fingerprinting combining features of on-board sensors. In: IEEE Trans Inf Forensics Secur 12:2457–2466
9. Alles EJ, Geradts ZJ, Veenman CJ (2008) Source camera identification for low resolution heavily compressed images In: International conference on computational sciences and its applications, 2008. ICCSA'08. IEEE, pp 557–567
10. Das A, Borisov N, Caesar M (2014) Do you hear what I hear?: fingerprinting smart devices through embedded acoustic components. In: Proceedings of the 2014 ACM SIGSAC conference on computer and communications security. ACM, pp 441–452
11. Kirchner M, Böhme R (2009) Synthesis of color filter array pattern in digital images. In: Media forensics and security, vol 7254. International Society for Optics and Photonics, p 72540K
12. Lukas J, Fridrich J, Goljan M (2006) Digital camera identification from sensor pattern noise. IEEE Trans Inf Forensics Secur 1(2):205–214
13. Lin X, Li C-T (2016) Preprocessing reference sensor pattern noise via spectrum equalization. IEEE Trans Inf Forensics Secur 11(1):126–140
14. Taspinar S, Mohanty M, Memon N (2017) PRNU-based camera attribution from multiple seam-carved images. IEEE Trans Inf Forensics Secur 12(12):3065–3080
15. Lukáš J, Fridrich J, Goljan M (2005) Determining digital image origin using sensor imperfections. In: Proceedings of SPIE electronic imaging, image and video communication and processing, vol 5685, pp 249–260
16. Lin X (2016) Digital image forensics based on sensor pattern noise. PhD thesis, University of Warwick
17. Yager N, Amin A (2004) Fingerprint classification: a review. Pattern Anal Appl 7(1):77–93
18. Lin X, Li C-T (2017) Large-scale image clustering based on camera fingerprints. IEEE Trans Inf Forensics Secur 12(4):793–808
19. Bertini F, Sharma R, Iannì A, Montesi D (2015) Profile resolution across multilayer networks through smartphone camera fingerprint. In: Proceedings of the 19th international database engineering & applications symposium. ACM, pp 23–32
20. Bertini F, Sharma R, Iannì A, Montesi D (2015) Smartphone verification and user profiles linking across social networks by camera fingerprinting. In: International conference on digital forensics and cyber crime. Springer, pp 176–186
21. Rouhi R, Bertini F, Montesi D (2018) A cluster-based approach of smartphone camera fingerprint for user profiles resolution within social network. In: Proceedings of the 22nd international database engineering & applications symposium. ACM, pp 287–291
22. Rouhi R, Bertini F, Montesi D, Li C-T (2019) Social network forensics through smartphones and shared images. In: 2019 7th international workshop on biometrics and forensics (IWBF). IEEE, pp 1–6
23. Shullani D, Fontani M, Iuliani M, Al Shaya O, Piva A (2017) Vision: a video and image dataset for source identification. EURASIP J Inf Secur 2017(1):15
24. Lin X, Li C-T (2018) Rotation-invariant binary representation of sensor pattern noise for source-oriented image and video clustering. In: 2018 15th IEEE international conference on advanced video and signal based surveillance (AVSS). IEEE, pp 1–6
25. Chen M, Fridrich J, Goljan M, Lukás J (2008) Determining image origin and integrity using sensor noise. IEEE Trans Inf Forensics Secur 3(1):74–90
26. Dabov K, Foi A, Katkovnik V, Egiazarian K (2007) Image denoising by sparse 3-d transform-domain collaborative filtering. IEEE Trans Image Process 16(8):2080–2095

27. Rouhi R, Bertini F, Montesi D, Lin X, Quan Y, Li C-T (2019) Hybrid clustering of shared images on social networks for digital forensics. IEEE Access 7:87288–87302
28. Carlson RE, Fritsch FN (1989) An algorithm for monotone piecewise bicubic interpolation. SIAM J Numer Anal 26(1):230–238
29. Schwenker F, Trentin E (2014) Pattern classification and clustering: a review of partially supervised learning approaches. Pattern Recogn Lett 37:4–14
30. Fahad A, Alshatri N, Tari Z, Alamri A, Khalil I, Zomaya AY, Foufou S, Bouras A (2014) A survey of clustering algorithms for big data: taxonomy and empirical analysis. IEEE Trans Emerg Top Comput 2(3):267–279
31. Li C-T, Lin X (2017) A fast source-oriented image clustering method for digital forensics. EURASIP J Image Video Process 2017(1):69
32. Ertöz L, Steinbach M, Kumar V (2003) Finding clusters of different sizes, shapes, and densities in noisy, high dimensional data. In: Proceedings of the 2003 SIAM international conference on data mining. SIAM, pp 47–58
33. Shirkhorshidi AS, Aghabozorgi S, Wah TY, Herawan T (2014) Big data clustering: a review. In: International conference on computational science and its applications. Springer, pp 707–720
34. Lloyd S (1982) Least squares quantization in PCM. IEEE Trans Inf Theory 28(2):129–137
35. Madhulatha TS (2011) Comparison between k-means and k-medoids clustering algorithms. In: International Conference on Advances in Computing and Information Technology. Springer, Berlin, Heidelberg. pp 472–481
36. Macqueen J (1967) Some methods for classification and analysis of multivariate observations. In: In 5-th Berkeley symposium on mathematical statistics and probability, pp 281–297
37. Park H-S, Jun C-H (2009) A simple and fast algorithm for k-medoids clustering. Expert Syst Appl 36(2):3336–3341
38. Kaufman L, Rousseeuw PJ (2009) Finding groups in data: an introduction to cluster analysis, vol 344. John Wiley & Sons, Hoboken, New Jersey
39. Refaeilzadeh P, Tang L, Liu H (2009) Cross validation, encyclopedia of database systems (EDBS). Arizona State University, Springer, p 6
40. Hubert L, Arabie P (1985) Comparing partitions. J Classif 2:193–218

Presentation Attacks in Mobile and Continuous Behavioral Biometric Systems

Tempestt Neal and Damon Woodard

Abstract Active authentication allows an individual's identity to be continuously verified in a transparent fashion. For devices centered on user convenience, active authentication using behavioral biometrics is an appealing solution to user authentication since behavioral data can be captured as consumers naturally interact with their devices. However, while such implementations are user-friendly and help to counter some of the challenges associated with knowledge-based authentication methods (e.g., easily guessed passcodes), an adversarial attack must be carefully considered. In this regard, to gain unauthorized access to a secured device, an adversary may falsify biometric information through impersonating the legitimate user. This attack is often referred to as a presentation attack or biometric spoofing. Throughout this chapter, various attack scenarios on mobile devices are discussed for gait, keystroke and touch dynamics, and user-device interaction modalities. Presentation attacks are categorized according to the biometric modality, which may differ given the context of the sensor component involved. This chapter exposes multiple research gaps and challenges which could significantly strengthen adversary detection once addressed, while discussing novel research in which no sensor information is required.

Keywords Behavioral biometrics · Continuous authentication · Mobile biometrics · Presentation attacks · Spoofing

---

T. Neal (✉)
Department of Computer Science and Engineering, University of South Florida, Tampa, FL, USA
e-mail: tjneal@usf.edu

D. Woodard
Department of Electrical and Computer Engineering, University of Florida, Gainesville, FL, USA
e-mail: dwoodard@ece.ufl.edu

T. Bourlai et al. (eds.), *Securing Social Identity in Mobile Platforms*, Advanced Sciences and Technologies for Security Applications,
https://doi.org/10.1007/978-3-030-39489-9_2

1 Introduction

Biometric recognition involves a methodical and (usually) fully automated process for establishing or authenticating a person's identity through the use of their physical and behavioral attributes. When establishing an identity, or performing the task of *identification*, a person's biometric data, such as fingerprints or gait signals, are extracted, decomposed into a digitized representation known as *features*, and compared to the features acquired from n individuals already known to the biometric system. Each known individual is said to be *enrolled*, and their enrolled features collectively form a set of *templates*. The features in question, or *query*, are matched against each template (i.e., one-to-many comparisons) to determine the most likely candidate. On the contrary, authentication requires a single comparison between a query and template (i.e., one-to-one comparison) due to the claiming of a template [29]. Thus, authentication ensures that a person claiming access to resources has the right to do so. In this chapter, biometric authentication on mobile devices is expanded upon, with a particular emphasis on the vulnerability of these approaches to adversarial attacks.

The use of biometric technology on mobile devices has surged over the past decade. However, many devices continue to rely on the knowledge of the user (e.g., passwords, personal identification numbers, or patterns) for access control [3, 4, 49, 61, 71, 73]. Hence, these methods are collectively referred to as *knowledge-based* authentication schemes. Some argue, however, that knowledge-based authentication is growing increasingly inefficient due to dictionary, shoulder surfing, and smudge attacks, the lack of use, and poorly chosen password combinations [13, 33, 35, 65]. On the other hand, since biometrics provide an automated means to person recognition based on *who* a person is instead of *what* a person knows, biometric authentication has gained a significant amount of commercial attention for user authentication. Correspondingly, physical biometric solutions, including face, fingerprint, and iris recognition, are becoming increasingly popular due to commercialization, convenience, and a perceived increase in robustness at reducing unauthorized device access [2, 5, 23, 66]. Unfortunately, physical biometric characteristics are also associated with a wide range of concerns. For example, they may degrade when the mobility of the device results in undesirable data acquisition conditions, and they may also be less conducive to active authentication (though not impossible [57]).

Active authentication involves the continuous acquisition of data; thus, a person's identity is continuously verified. The advantages associated with active authentication are substantial. Since active authentication involves unobtrusive and uninterrupted monitoring, individuals are not burdened with the authentication process. Hence, active authentication may be regarded as more user-friendly and secure compared to knowledge-based and point-of-entry biometric authentication methods such as those currently used commercially. Since the latter does not provide session-long protection, the application and its data, services, and other relevant artifacts are available to whomever until the authorized session ends (e.g., the device locks automatically or manually). Further, there is never a guarantee, especially with knowledge-based authentication, that the person claiming access to a device is indeed an authorized individual. On the other hand, since biometric recognition relies upon intrinsic data and active authentication continuously captures this data, the combination is much more vigilant, and the output of many authentication decisions over time is intended to reflect a higher level of awareness of a person's identity.

Behavioral characteristics may be more appropriate for supporting active authentication. Behavioral biometrics seek to decouple the human and device in the authentication process through passive and covert acquisition of data. Therefore, they are less dependent on hardware, less constrained by external environmental factors such as lighting conditions, and can be continuously and efficiently captured as users naturally interact with their device. Nonetheless, as a whole, biometric systems are not unerring; in fact, biometric errors and attacks are both thriving research topics [20, 21, 56]. Physical biometrics have a long history of being susceptible to presentation attacks, or spoofing (i.e., an attacker presents false data that appears to belong to a legitimate subject to bypass liveness detection schemes and ultimately return an inaccurate matching decision), and reliance upon ubiquitous sensors in uncontrolled and unattended environments such as accelerometers or fingerprint scanners on mobile devices increases the risk of unauthorized access. Behavioral biometrics are also susceptible to spoofing, especially through observation and imitation. Threats against biometric systems are depicted in Fig. 1 and are defined below.

1. **Presentation/spoofing attack**: A fake biometric (e.g., fingerprint mold or synthetic irides [19, 67]) is presented to the sensor to appear as legitimate data.
2. **Replay attack**: Previously seen biometric data is resubmitted, or replayed, to the feature extraction module; the sensing module is bypassed (i.e., no query data is acquired) [28].
3. **Override the feature extractor**: The feature extraction algorithm is modified to extract features of the attacker's choice.
4. **Template theft or modification**: Stored data may be modified to allow an attacker to appear as if (s)he is enrolled or to prevent access to a legitimate person.
5. **Intercept communication channels**: The channels connecting the feature extractor to the database, feature extractor to the matcher, and database to the matcher may be intercepted. For example, packet sniffers may be an effective means for intercepting channels if preventative measures are not in place, such as the Simple Network Management Protocol [11].
6. **Override the matcher**: The matching scheme is modified to produce a matching score set/computed by the attacker.
7. **Override the decision**: The authentication decision (i.e., genuine or impostor) is changed to suit the needs of the attacker [54].

On mobile devices, sensing components are much easier to access compared to those found in traditional biometric systems. Face, fingerprint, and iris data rely on *uncontrolled* and *unattended* cameras, lighting sources, and scanners. Thus,

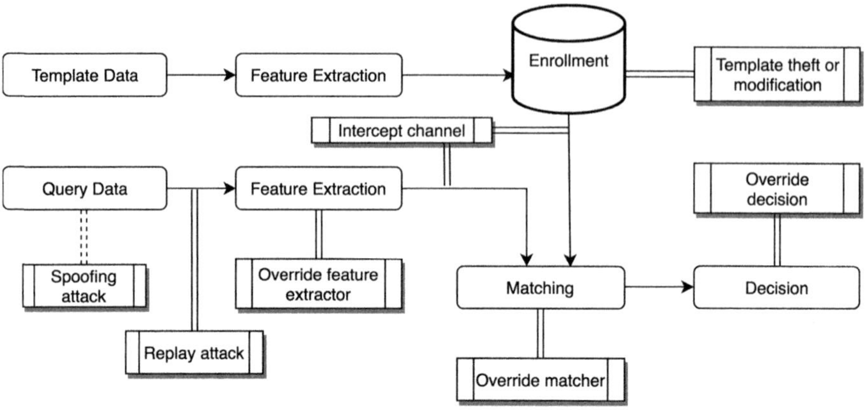

Fig. 1 Points of attack indicated by double lines in a biometric system

manufacturers tend to give their consumers more control, trust, and flexibility with the biometric service with little protection against sensor tampering. Furthermore, as shown in Fig. 1, biometric recognition is a delicate process; the flow of data from a person to an identification decision is permeated with many points of vulnerability. Beyond the *direct* entry point where presentation attacks are possible, *indirect* attacks such as template theft and channel interception are also plausible. Fortunately, template protection schemes have been combined with various protocols and specifications (e.g., The Fast Identity Online Alliance, Biometric Open Protocol Standard, and Trusted Execution Environments) to minimize indirect attack success [27, 28, 37, 51, 64, 74, 76]. However, spoofing remains a grave concern considering easy-to-access sensors and the minimal amount of knowledge required by an attacker regarding the internal configuration of the biometric system to launch a successful presentation attack [53]. Due to a growing interest in active authentication over the past few years (e.g., [62, 68]), it is important to consider not only the practical potential of active authentication, but also theoretical and practical implications regarding the vulnerability of such systems. This chapter highlights many of these challenges and the current state of academic research.

1.1 Biometric Performance

The performance of a biometric system is measured by its likelihood to return an incorrect decision (i.e., a legitimate, or genuine, subject is falsely rejected or an attacker, or impostor subject, is falsely accepted). Generally, a threshold value, t, determines the similarity between a template and query. If the matching scores are generated based on a similarity-based metric, then matching scores exceeding t indicate a match, while scores less than t indicate a non-match. Likewise, if scores are produced based on the distance between the template and query, matching scores below t indicate a match, while scores exceeding t indicate a non-match. Assuming a distance-based matcher, as t decreases, the proportion of rejected genuine subjects increases. As t increases, the proportion of accepted impostor subjects increases. Thus, the threshold at which these two values are approximately equal, or the equal error rate (EER), is usually sought after as the optimal operating point. Other common measures of performance include:

- A **true positive** is the correct prediction of a genuine subject as genuine.
- A **true negative** is the correct prediction of an impostor subject as an impostor.
- A **false positive** is the incorrect prediction of an impostor as genuine.
- A **false negative** is the incorrect prediction of a genuine subject as an impostor.
- The **true positive rate** (TPR) is the proportion of true positives to all positives $(TP/TP + FN)$.
- The **true negative rate** (TNR) is the proportion of true negatives to all negatives $(TN/TN + FP)$.
- The **false positive rate** (FPR) (also referred to as the *false accept rate* (FAR) or *impostor pass rate* (IPR)) is the proportion of false positives to all negatives $(FP/FP + TN)$.
- The **false negative rate** (also referred to as the *false reject rate* (FRR)) is the proportion of false negatives to all positives $(FN/TP + FN)$.

Successful presentation attacks increase the false positive rate, which in turn increases the equal error rate. Many successful presentation attacks result in the shifting of impostor scores closer to the distribution of genuine scores; thus, the false rejection rate increases due to the increased amount of overlap between the two distributions (Figs. 2 and 3). Thus, since mobile devices are used for multiple purposes such as monitoring and managing social networking sites, composing and receiving e-mails and other forms of communication, e-banking, and capturing and storing various forms of media, this chapter focuses on a timely security issue facing mobile biometrics. Up to now, the mobile device market has relied on physical biometrics, and many people are habituated with these services. Meanwhile, studies have shown that individuals tend to adopt methods which are most familiar [77], and thanks to the influence of mobile biometrics, biometric technology is no longer unusual. At the time of this writing, a simple Google Scholar search of 'continuous authentication' returns over 19,000 publications, patents, and citations all published from 2018. The adoption of behavioral biometrics on mobile devices for active authentication may not be as far-fetched as it seems, and perhaps should be expected rather than speculated. Thus, this chapter highlights the state of presentation attack detection on mobile platforms assuming an active authentication approach. Section 2 describes gait recognition and the accompanying academic literature which explores the effectiveness of imitating a person's walking patterns. Section 3 focuses on keystroke dynamics and touch gestures, and the many works which consider the theft of keystroke data through remote spoofing attacks. Section 4 discusses the extraction of user-device interactions (e.g., mobile application usage) as behavioral biometric data and how these data may be spoofed in the absence

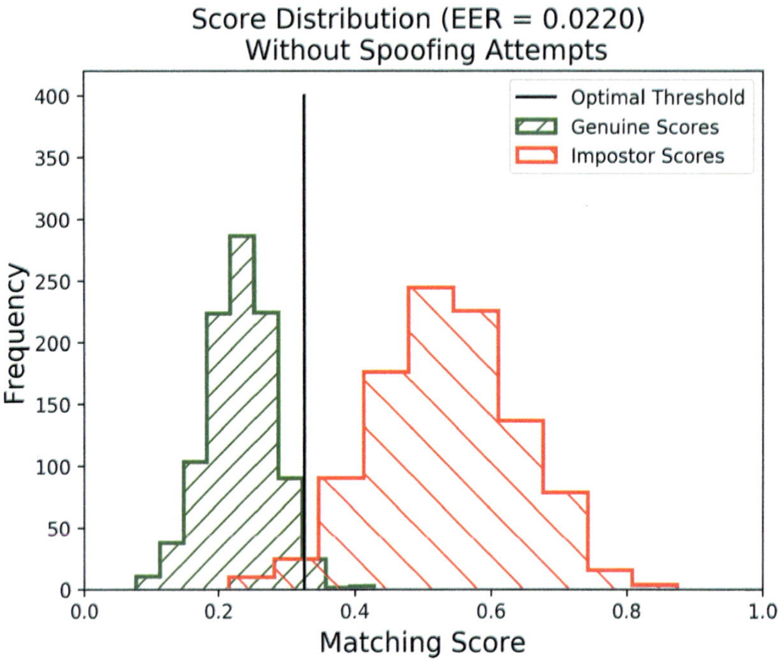

Fig. 2 Example score distribution assuming no spoofing attempts were made

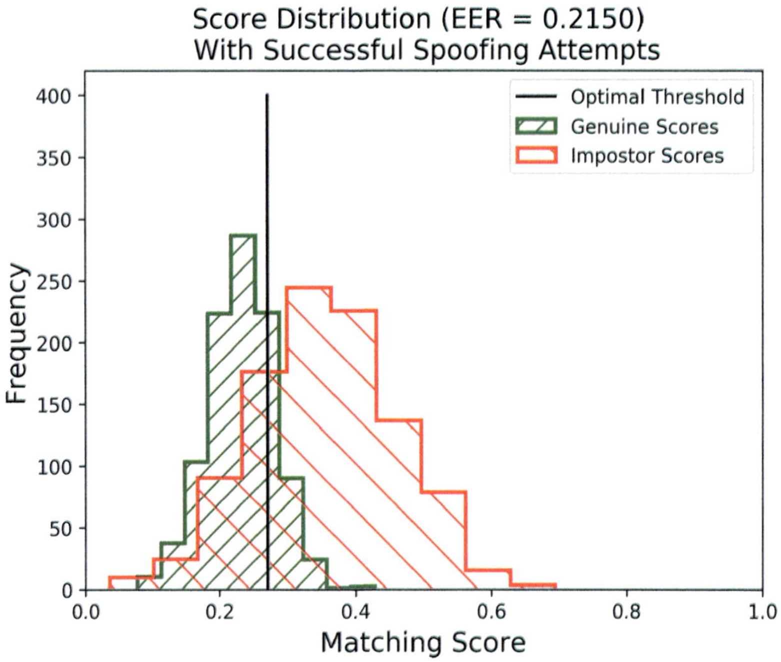

Fig. 3 Example score distribution assuming several spoofing attempts were made and were successful

of specialized sensors. Remaining challenges and considerations are provided in Sect. 5, with a chapter summary provided in Sect. 6.

2 Gait

Gait recognition, or the systematic process of extracting and utilizing a person's walking patterns for authentication, has prevailed as a biometric modality for at least twenty years [48]. There are three main methods for acquiring gait data: video recordings, floor sensors, and wearable sensors (WS) [75]. In each of these, data can be acquired covertly with little to no user cooperation. However, the latter approach is arguably most suitable for continuous authentication since the sensing mechanism is attached to an individual and mobile devices have embedded inertial sensors (e.g., accelerometers and gyroscopes) which can capture walking patterns [47]. Since many people carry their smartphones in a pocket, for instance, these sensors capture rich and idiosyncratic movement signals [15].

The process of gait recognition involves five main steps: sampling, noise reduction, cycle detection, feature extraction, and matching. Sampling is simply data acquisition and there a variety of approaches explored in the academic literature in terms of what type of sensing device is used (e.g., variations in accelerometer hardware models) and where the device is placed (e.g., in a shirt pocket or on a belt clip). Raw walking signals, which are usually three-dimensional accelerometer measurements (Fig. 4), correspond with a variable number of steps per minute infiltrated with unwanted noise. Inertial sensors are notorious for capturing spurious

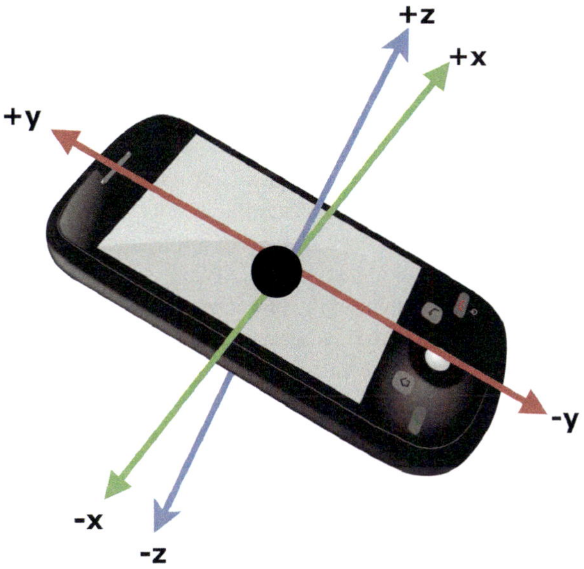

Fig. 4 Three-dimensional measurement of movement via accelerometers on a mobile device

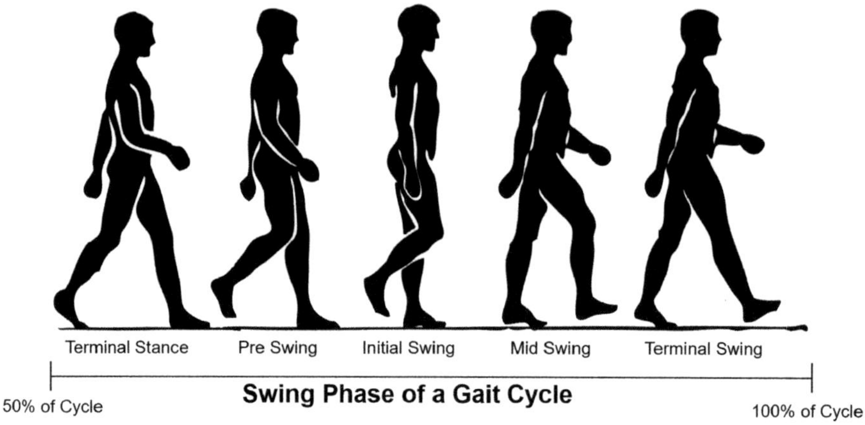

Fig. 5 Swing Phase of a Gait Cycle

movements, such as those resulting from breathing or inadvertent device movements in a pocket. Linear or cubic interpolation is often used to create equal time intervals between each data point, while smoothing methods, such as weighted moving averaging, are also employed. The enhanced data is then used for cycle detection.

One cycle consists of two steps (Fig. 5); therefore, a walk is composed of several cycles. There are various means to cycle detection; a common approach is to utilize local minima and maxima in the enhanced data. Once each cycle is found, the average cycle is computed and stored as a unique representation of a person's gait. Matching is thus performed between two average gait cycles.

There are a number of studies that consider presentation attacks in gait biometrics in the context of video recordings [8, 25, 26, 38]. These efforts highlight the risks of "clothing impersonation" [8]. However, muscular and skeletal attributes play a larger role in WS approaches since visual information is not available. Presentation attack success in WS approaches are therefore dependent on an attacker's ability to somehow adopt the musculoskeletal singularities of their target [16].

Two main presentation attacks on gait biometric systems have been explored: passive and active. Passive spoofing attacks require minimal effort from the attacker; the attacker has not been trained to mimic their target, nor has the attacker studied their target's walking style [16]. Passive attacks are commonly known as *zero-effort* attacks throughout the biometric literature, where data from another randomly chosen subject is used as attack data. Hence, the research literature refers to this scenario as *friendly*. An active attack occurs when an attacker purposely tries to match their target or select the most suitable person to target; appropriately, these attacks create a *hostile* scenario. For example, Gafurov, Snekkenes, and Bours used a motion-recording sensor attached to the hip to collect data at a rate of 100 samples per second to analyze passive and active presentation attacks [16, 17]. The magnitude of the acceleration vectors, $\sqrt{x_i^2 + y_i^2 + z_i^2}, i = 1, \ldots, k$, was used to combine the data from all directions into a single signal. The resulting signal was linearly interpolated and smoothed with a moving average filter. Cycle detection involved searching for the minimal acceleration points, and average cycles were compared using the Euclidean distance. One hundred subjects participated in the friendly scenario, and 90 participated in the hostile scenario. In the latter, attackers and targets were paired according to similarities in gender, age, height, and weight, and having known each other prior to the study. While the EER for the friendly scenario was 13%, a *closest target* attack, or a hostile attack where the attacker chooses a target with similar walking patterns as their own, significantly increased the FAR.

Searching for minima for cycle detection may lead to error, however, when the first cycle is inaccurate (and thus, each cycle thereafter is also inaccurate). Ren et al. utilized correlation for cycle detection as an alternative method [55]. Interestingly, the authors found that a person's cycles are highly correlated regardless of walking speed. By placing HTC smartphones on the hip of 23 subjects, the authors found that walks of at least 20 seconds maximized the attack detection rate in passive scenarios. Thus, their results suggest that the amount of data has some bearing on the ability to distinguish between genuine and impostor subjects.

Correlation-based gait recognition is not an uncommon approach. For instance, a more aggressive attacker was assessed by Mjaaland et al. [41]. Attackers were trained to mimic their targets by viewing video recordings, having access to statistical data, and coaching. In this effort, dynamic time warping was used for matching accelerometer readings since it copes with changes in speed. A motion recording device was attached to the hip for capturing 100 samples per second. Correlation between maxima in the walking signals was computed for estimating cycle length, simplifying cycle detection. Two versions of hostile attackers were analyzed; a short-term active attacker was trained by watching videos of their victim's walking style for an hour, while a long-term active attacker was trained for six weeks. The goal of this study was to analyze the attackers' abilities to learn over time. A noticeable result of this study regarding spoofing abilities was the identification of a *plateau*, or the point at which an attacker can no longer improve his or her ability to mimic their target; they mathematically defined this point as $p = \lim_{x\to\infty} Y(X)$, where $Y(X)$ was an observation at instance X [39]. Many attackers worsened their performance during training, but a few were able to improve temporarily. One of the improving attackers learned to focus on certain characteristics of his gait for improvement in mimicking their target, but expressed frustration from the feeling of walking in an unnatural way. Another attacker which improved made significant changes to his gait to do so, but also began to walk unnaturally. Thus, the authors claimed that "if exactly one plateau exists for each individual, then the success of an attacker is predetermined – the plateau has to lie below or near the acceptance threshold for an impostor to ever be able to succeed" [40].

In general, the research literature suggests that spoofing gait biometrics is not an easy task. For example, work by Muaaz and Mayrhofer also demonstrates the difficulty of spoofing gait biometric systems [43]. They analyzed two forms of presentation attacks: reenact and coincide. In the former, the attacker walked behind their victim during rehearsal, and then mimicked alone during reenactment. In the coincide method, the victim and attacker walked side by side. For both, victim and attacker were instructed to wear similar clothing. Moreover, attackers were carefully chosen; they were trained mime artists and thus specialized in copying body motions and language. They were also similar to their victims in age, weight, height, shoe size, and leg length. However, none of the attackers created a false positive during reenactment or coincide. Thus, one may question whether the literature has thoroughly considered the characteristics that make a target vulnerable.

There are a few additional open challenges associated with gait biometrics and presentation attacks. As mentioned, several research efforts focus on visual-based gait recognition (i.e., video recordings) and the risk of spoofing due to similarity in items such as clothing. Future research should place more emphasis on WS-based gait biometrics and active and passive attackers. Second, existing literature is unfortunately limited in experimental conditions. Typically, laboratory settings are used such that certain parameters are controlled and external noise is minimized. For instance, it is common for gait recognition studies to restrict where the sensing device is placed (e.g., on a belt clip) and the walking surface of the subjects (e.g., a hallway). However, this may lead to some discrepancy between experimental and actual walking conditions. Another consideration is the trade-off between minimizing attack detection time and the amount of time given for data capture. When only a few steps are used for training (i.e., cycle detection), an individual's natural walking pattern may not fully form. Meanwhile, an attacker is not given enough time for observation; therefore, both false positives and false negatives may increase [41]. Nonetheless, gait biometrics is more than feasible on mobile devices, and they support the continuous and passive nature of active authentication; perhaps the future of gait recognition will lead to commercialized authentication solutions given the current indication that spoofing such systems is a laborious task for attackers.

3 Keystroke Dynamics and Touch Gestures

Keystroke dynamics were first explored on traditional QWERTY keyboards (top of Fig. 6), where features such as key hold latency (time between press and release of the same key), key press latency (time between press of a key and the next key), and interkey latency (time between release of a key and press of the next key) were extracted. As the design of mobile devices evolved through the years from mini 3D QWERTY keyboards to touch screens (bottom of Fig. 6), keystroke dynamics and touch gestures emerged as mobile behavioral biometric modalities. Touch gestures, such as zooming, scrolling, and swiping (Fig. 7), are characterized using features such as the starting position of the touch, finger pressure and area, and gesture direction, distance, duration, and curvature [34].

There are two common themes in the academic literature concerning spoofing of keystroke and touch data: the assumption of a generative, non-physical attack and

Fig. 6 Variations in keyboards which have been explored in keystroke dynamics research

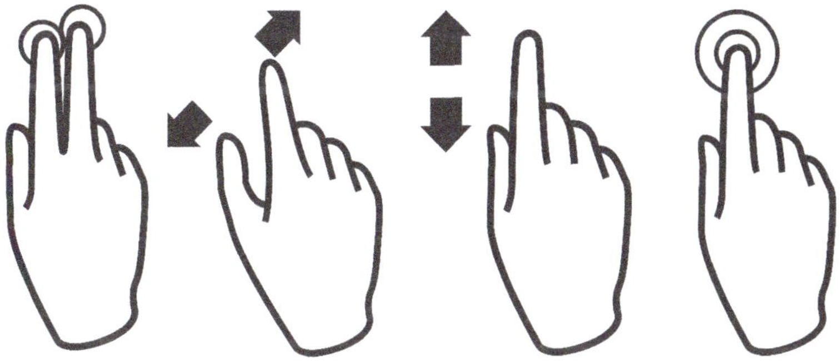

Fig. 7 Examples of two-finger, zoom, scroll, and double-press touch gestures, respectively

the analysis of such attacks on desktop computers. For example, Stefan, Shu, and Yao explored an attacker that can successfully create a program to gather statistical information regarding legitimate keystrokes to subsequently reproduce similar, but spoofed, data [63]. Client-side malware would then inject false keystroke information. Thus, their attack scenario assumes the attacker can generate statistically similar keystroke data from a remote location by implementing a keylogger to obtain actual data. Since several hardware and software keyloggers exist which can covertly obtain keystroke data, while emulators (i.e., programs that can generate keystroke and touch events) are easy to develop, remote presentation attacks are a concern for these modalities.

Rahman et al. also examined remote attacks; they referred to their model as a *snoop-forge-replay* attack since the attacker snoops on their target's data, forges a similar copy, and replays the snooped data [52, 53]. They made the following observations:

- The probability of error significantly increased during an attack; EERs increased from 11% to over 58%.
- Spoofing was most successful when the forged copy of data was generated from a small amount of legitimate data; the authors attributed this observation to the heavy-tailed distribution of digraphs in the English language.
- Legacy keystrokes, or those occurring six months after training, were at risk of spoofing.
- Shorter keystroke samples were also at risk of spoofing, but the forged samples were not linguistically sound.

Despite the insights gained from these efforts, it remains unclear as to whether these observations apply to mobile platforms. For example, in an active authentication setting on mobile devices, legacy keystrokes may be less susceptible to spoofing since keystroke features may change according to the location of the user. In addition, a large amount of keystroke information may be available on desktop computers, while keystroke data from mobile devices may be constrained to only a few characters. Responses from key presses (3D button versus haptic feedback) may also alter a person's keystroke and touch input. Finally, because the literature has focused on generative attacks (i.e., the attacker computes the statistical likelihood of certain features to generate accurate spoofed data), an attacker is assumed to have some knowledge regarding the likelihood of certain features, and uses this knowledge to simulate keystroke activity [58, 59]. On mobile platforms, however, keystroke and touch data may be more erratic; features may vary according to task (e.g., gaming versus e-banking), while touch data has not been explored in its entirety in regard to generative presentation attacks.

Nonetheless, there is one particular publication which has focused on active authentication on mobile devices and presentation attacks [31]. Three popular touch gesture schemes were evaluated for their robustness against spoofing: SilentSense [7], Touchanalytics [14], and Li et al.'s method [34]. Fifty-five participants were recruited from which touch data was collected along with multi-angled videos of nine of the subjects. Thirty-two attackers were instructed on how to execute shoulder surfing and offline training attacks. In the former, the attackers studied the video recordings of the legitimate subjects; in the latter, a mobile application was used to train the attackers on how to reproduce spoofed samples by providing guidance such as "move start point towards right." Attackers were incredibly successful in these experiments, with attack success rates averaging 81.7% and 84.7% for shoulder surfing and offline training, respectively. Moreover, attackers required very little effort before gaining a sufficient amount of knowledge to succeed; for 15% of successful shoulder surfing attacks, attackers only needed 30 seconds of video observation, while the majority required less than two minutes. The authors also found a negative correlation between the number of retries during shoulder surfing attacks and attack success rate, indicating that attackers were not required to obtain a certain level of skill. Thus, while many works focus on generative attacks, it is clear that actual presentation attacks through simple observation may also pose great risk to mobile device users.

4 User-Device Interaction

An interesting way to support active authentication on mobile devices is to allow the device to "get to know its owner" [6]. In such scenarios, emphasis is placed on how the user interacts with their device; thus, explicit use of sensor information is not a critical factor for data collection [44]. Instead, mobile devices are equipped with several software-based services (i.e., mobile applications), and studying how a person chooses to interact with these services (e.g., when a mobile application is launched or how much time a person spends on an application) forms the basis for user-device biometric data. In regard to spoofing these types of biometric systems, there are various approaches that have been considered. There is a common theme, however, to consider levels of attackers. *Informed* adversaries are knowledgeable of some aspect of their target's behavior (e.g., web browsing habits), while an *uninformed* adversary is similar to a zero-effort or passive attacker. For example, Neal et al. simulated presentation attacks derived from actual activities of legitimate subjects [45]. They extracted association rules from application, Bluetooth, and Wi-Fi activity logs as features from 189 subjects and implemented four levels of informed adversaries. The first level was generated by creating a threat model consisting of 50% legitimate data (from the attacker's target) and 50% noise. The second level reduced the amount of noise to 25%. All noise was removed in the third level, while the fourth level was a near replication (with some discrepancies) of the target's features. Using the Jaccard distance for matching, they found that all subjects were susceptible to the third and fourth level attacks. They implemented the second level threat model in an active authentication protocol, where EERs reached up to 44% [46].

Meanwhile, Bicakci et al. presented Device Comfort as an anomaly detection application [6]. Device Comfort's functionality focused on contextualizing application, Bluetooth, calling, text messaging, and Wi-Fi activities for computing a anomaly score using the Reality Mining Dataset [12] (100 subjects) and the Social Evolution Dataset [36] (80 subjects) for experimentation. The researchers focused their efforts on diurnal patterns, where n days of data was used for training and data following midnight of the last training data was used for testing. Their approach used a variant of the k-nearest neighbors algorithm; samples with far or few neighbors were considered anomalies. Using this methodology, their findings indicated that certain data types were more informative than others. For instance, location, call history, and Bluetooth connectivity were most useful for the Social Evolution dataset.

It is important to note that the few studies which have focused on user-device interaction and presentation attacks tend to isolate certain elements of their experimental designs and subject these elements to spoofing. For example, as mentioned, Neal et al. considered four levels of spoofing independently, while Bicakci et al. assumed 'daily' spoofing attempts. In additional efforts, the 'goodness' and 'badness' of activities were analyzed, where a good event was associated with a familiar action (such as calling a number in the phonebook), while a bad event was unfamiliar [30, 60]. Similarly, Yazji et al. focused on spatio-temporal activities, or users' locations at particular times using area IDs (e.g., a library or office) [72]. Thus, there is ample room on the topic of user-device interactions for exploring presentation attacks. Since this particular modality is sensor-independent, it may be the case that an attacker would have to dedicate a significant amount of time to observation. On the other hand, it could be possible that generative attacks, such as those explored in keystroke dynamics, are more plausible since it would then be less likely that an attacker would be noticed by their target. Nonetheless, both presentation attack techniques deserve more attention for the derivation of effective anti-spoofing measures.

5 Open Challenges

Although this chapter discusses several research efforts, many open challenges remain. Our discussion shows that an emphasis is placed on generative attacks in the research literature, which assumes that an attacker has prior knowledge of the feature representation. Thus, choice of features is an important component of presentation attack detection. There are also very few (if any) experiments ran in the wild, while the data available to the research community is significantly limited. This section elaborates on some of these challenges.

Importance of Features A presentation attack should be most successful when the falsified biometric data yields relevant and statistically similar features to an enrolled victim. Otherwise, attacks will likely be detected (assuming the attacker's data is not like that of someone other than the victim). This fact places considerable importance on the feature extraction module, and consequently, feature generation. Thus, feature generation should entail the process of ensuring an answer of *very little* to the question of *how much can an attacker learn about their target victim*, especially when considering generative presentation attacks.

Feature Representations People tend to use mobile devices in a habitual manner due to many internal (e.g., neurological) and external (e.g., contextual) factors [10, 50]. Thus, human behavior is quite obscure, creating a rather unconstrained pattern recognition problem in terms of a biometric system [32, 70]. Yet, many research efforts have utilized standard sets of hand-crafted features (e.g., keystroke timings), which may fail to accurately capture the complexity of behavioral biometric modalities. Increasing the number of feature dimensions may create more robust feature representations that decrease intra-person variance while increasing inter-person (including attackers) separation. Since attackers would then be required to generate a larger set of features, the task of accurately understanding the statistical nature of legitimate features may become much harder. Another possible solution is to utilize person-dependent feature selection, such that a standard feature set can no longer be exploited by attackers, along with learned features through deep learning.

Error and Liveness Detection Physically detecting presentation attacks may be more feasible when an actual human has obtained unauthorized possession of a mobile device. However, as mentioned, many presentation attacks explored in the research literature are generative attacks that occur remotely (e.g., the snoop-forge-replay keystroke attack [52]). However, in one particular study of keystroke dynamics, negative keystroke timing features were observed among legitimate subjects [63]. The authors of this work suggested that it may be difficult for a bot to replicate this behavior. Further, these features (i.e., negative inter-key timing and duration) were the most salient. Since negative durations were only present for longer texts, it would be interesting to observe how these findings translate to mobile devices. If such claims generalize well to multiple platforms, it may be worth considering the inclusion of features which focus exclusively on human error as a means to detecting and preventing remote presentation attacks.

Assessment of Actual Attacks The research community is unfortunately limited in experimental environments. This may lead to discrepancy between actual environments that are (or are not) conducive to presentation attacks and those used during experimentation. For example, in a study of gait recognition, attackers are trained to mimic their targets, but all experiments were conducted indoors on the same flooring [41]. In many studies on keystroke dynamics, subjects provide input on desktop computers with a traditional keyboard [42]. Thus, there may exist a significant gap between theory and practice.

Lack of Benchmark Datasets There is a significant lack of benchmark datasets. As a result, many efforts cannot be replicated, while many research challenges remain. For instance, a dataset which consists of several types of subjects (e.g., variations in age, ethnicity, and employment) would aid in determining the characteristics of people that make them more or less vulnerable to certain spoofing attacks.

6 Summary

Biometric recognition has gained a significant amount of attention for user authentication on mobile devices. Presentation, or spoofing, attacks, where an attacker falsifies data to appear legitimate, are one of several threats against biometric systems [18, 24, 69]. Presentation attacks which occur during active authentication are interesting in their own right, and current research continues to focus on this problem. Since presentation attacks occur at the sensor level, they are quite frightening for mobile device users. These systems are unattended and uncontrolled, and these characteristics alone are alarming, especially considering the large population which interact with these systems.

This chapter discusses presentation attacks on mobile devices regarding gait, keystroke dynamics and touch gestures, and user-device interactions. These modalities were chosen since they are the most likely to be considered commercially for active authentication and have received a significant amount of attention in recent years from the research community. Further, current biometric systems on mobile devices rely on physical data, and manufacturers seem to have a solid grasp on implementing solutions for reliably detecting adverse situations for physical modalities.

Unfortunately, the research literature is minimally standardized on this topic; most efforts are not replicated, while many approaches are extremely controlled. However, since Apple's Touch ID was introduced in 2013 [1, 22], there has been a significant shift in the adoption of biometric technology. In fact, biometric technologies are so pervasive that they are becoming an important factor in the overall design of mobile devices [9]. Even though the mobile device market has relied on physical biometrics, active authentication has caught on as well over the past few years, and it could very well become a part of daily life in the near future. Thus, this chapter explores a timely research topic and an active research area; it is expected that presentation attack detection regarding behavioral biometric modalities will continue to improve over the next few years.

References

1. Use touch id on iphone and ipad. https://support.apple.com/en-us/HT201371
2. About touch id security on iphone and ipad (2015) https://support.apple.com/en-us/HT204587. Accessed 1 Sept 2018
3. Adams A, Sasse MA (1999) Users are not the enemy. Commun ACM 42(12):40–46. https://doi.org/10.1145/322796.322806
4. Bates TB (2016) Smartphone security: why doodling trumps text passwords. http://news.rutgers.edu/news/smartphone-security-why-doodling-trumps-text-passwords/20160309
5. Bhagavatula C, Ur B, Iacovino K, Kywe SM, Cranor LF, Savvides M (2015) Biometric authentication on iphone and android: usability, perceptions, and influences on adoption. In: Proceedings of USEC
6. Bicakci MV, Esfandiari B, Marsh S (2014) Anomaly detection for mobile device comfort. In: IFIP international conference on trust management. Springer, pp 93–108
7. Bo C, Zhang L, Li XY, Huang Q, Wang Y (2013) Silentsense: silent user identification via touch and movement behavioral biometrics. In: Proceedings of the 19th annual international conference on mobile computing and networking, MobiCom'13. ACM, New York, pp 187–190. https://doi.org/10.1145/2500423.2504572
8. Bustard JD, Ghahramani M, Carter JN, Hadid A, Nixon MS (2014) Gait anti-spoofing. Springer, London, pp 147–163. https://doi.org/10.1007/978-1-4471-6524-8_8

9. Byford S (2017) The first phone with an in-screen fingerprint sensor will come from vivo. https://www.theverge.com/circuitbreaker/2017/12/15/16779916/vivo-fingerprint-sensor-embedded-in-screen
10. van Deursen AJ, Bolle CL, Hegner SM, Kommers PA (2015) Modeling habitual and addictive smartphone behavior: the role of smartphone usage types, emotional intelligence, social stress, self-regulation, age, and gender. Comput Human Behav 45:411–420. https://doi.org/10.1016/j.chb.2014.12.039
11. Ding J (2016) Advances in network management. Auerbach Publications. Boca Raton, FL 33487 USA
12. Eagle N, (Sandy) Pentland A: Reality mining: sensing complex social systems. Pers Ubiquit Comput 10(4):255–268 (2006). https://doi.org/10.1007/s00779-005-0046-3
13. Egelman S, Jain S, Portnoff RS, Liao K, Consolvo S, Wagner D (2014) Are you ready to lock? In: Proceedings of the 2014 ACM SIGSAC conference on computer and communications security, CCS'14. ACM, New York, pp 750–761. https://doi.org/10.1145/2660267.2660273
14. Frank M, Biedert R, Ma E, Martinovic I, Song D (2013) Touchalytics: on the applicability of touchscreen input as a behavioral biometric for continuous authentication. IEEE Trans Inf Forensics Secur 8(1):136–148. https://doi.org/10.1109/TIFS.2012.2225048
15. Gafurov D, Snekkenes E (2009) Gait recognition using wearable motion recording sensors. EURASIP J Adv Signal Process 2009:7:1–7:16. https://doi.org/10.1155/2009/415817
16. Gafurov D, Snekkenes E, Bours P (2007) Spoof attacks on gait authentication system. IEEE Trans Inf Forensics Secur 2(3):491–502. https://doi.org/10.1109/TIFS.2007.902030
17. Gafurov D, Snekkenes E, Buvarp TE (2006) Robustness of biometric gait authentication against impersonation attack. In: Meersman R, Tari Z, Herrero P (eds) On the move to meaningful internet systems 2006: OTM 2006 workshops. Springer, Berlin/Heidelberg, pp 479–488
18. Galbally J, Marcel S, Fierrez J (2014) Biometric antispoofing methods: a survey in face recognition. IEEE Access 2:1530–1552. https://doi.org/10.1109/ACCESS.2014.2381273
19. Galbally J, Marcel S, Fierrez J (2014) Image quality assessment for fake biometric detection: application to iris, fingerprint, and face recognition. IEEE Trans Image Process 23(2):710–724. https://doi.org/10.1109/TIP.2013.2292332
20. Ghiani L, Marcialis GL, Roli F (2017) Fingerprint presentation attacks detection based on the user-specific effect. In: 2017 IEEE international joint conference on biometrics (IJCB), pp 352–358. https://doi.org/10.1109/BTAS.2017.8272717
21. Gomez-Barrero M (2018) Predicting the vulnerability of biometric systems to attacks based on morphed biometric information. IET Biom 7:333–341(8)
22. Goode A (2014) Bring your own finger – how mobile is bringing biometrics to consumers. Biom Technol Today 2014(5):5–9. https://doi.org/10.1016/S0969-4765(14)70088-8
23. Google: Try face unlock. https://support.google.com/nexus/answer/2781894?hl=en. Accessed 1 Sept 2018
24. Hadid A, Evans N, Marcel S, Fierrez J (2015) Biometrics systems under spoofing attack: an evaluation methodology and lessons learned. IEEE Signal Process Mag 32(5):20–30. https://doi.org/10.1109/MSP.2015.2437652
25. Hadid A, Ghahramani M, Bustard J, Nixon M (2013) Improving gait biometrics under spoofing attacks. In: Petrosino A (ed) Image analysis and processing – ICIAP 2013. Springer, Berlin/Heidelberg, pp 1–10
26. Hadid A, Ghahramani M, Kellokumpu V, Pietikäinen M, Bustard J, Nixon M (2012) Can gait biometrics be spoofed? In: Proceedings of the 21st international conference on pattern recognition (ICPR2012), pp 3280–3283
27. Holden W (2016) Securing public faith in biometrics. Biom Technol Today 2016(9):7–9
28. Jain AK, Nandakumar K, Nagar A (2008) Biometric template security. EURASIP J Adv Signal Process 2008:113:1–113:17. https://doi.org/10.1155/2008/579416
29. Jain AK, Ross AA, Nandakumar K (2011) Introduction to biometrics. Springer Science & Business Media, New York, NY 10013 USA

30. Jakobsson M, Shi E, Golle P, Chow R (2009) Implicit authentication for mobile devices. In: Proceedings of the 4th USENIX conference on hot topics in security. USENIX Association, p 9
31. Khan H, Hengartner U, Vogel D (2016) Targeted mimicry attacks on touch input based implicit authentication schemes. In: Proceedings of the 14th annual international conference on mobile systems, applications, and services, MobiSys'16. ACM, New York, pp 387–398. https://doi.org/10.1145/2906388.2906404
32. Kumar U, Kim J, Helmy A (2013) Changing patterns of mobile network (WLAN) usage: Smart-phones vs. laptops. In: 2013 9th international wireless communications and mobile computing conference (IWCMC), pp 1584–1589. https://doi.org/10.1109/IWCMC.2013.6583792
33. Kwon T, Na S (2014) Tinylock: affordable defense against smudge attacks on smartphone pattern lock systems. Comput Secur 42:137–150. https://doi.org/10.1016/j.cose.2013.12.001
34. Li L, Zhao X, Xue G (2013) Unobservable reauthentication for smartphones. In: In network and distributed system security symposium. The Internet Society
35. Li Z, Sun Q, Lian Y, Giusto DD (2005) An association-based graphical password design resistant to shoulder-surfing attack. In: 2005 IEEE international conference on multimedia and expo, pp 245–248. https://doi.org/10.1109/ICME.2005.1521406
36. Madan A, Cebrian M, Moturu S, Farrahi K, Pentland A (2012) Sensing the "health state" of a community. IEEE Pervasive Comput 11(4):36–45. https://doi.org/10.1109/MPRV.2011.79
37. Mandt T, Solnik M, Wang D (2016) Demystifying the secure enclave processor. Black Hat Las Vegas
38. Masood H, Farooq H (2017) A proposed framework for vision based gait biometric system against spoofing attacks. In: 2017 international conference on communication, computing and digital systems (C-CODE), pp 357–362. https://doi.org/10.1109/C-CODE.2017.7918957
39. Mjaaland BB (2009) Gait mimicking: attack resistance testing of gait authentication systems. Master's thesis, Norwegian University of Science and Technology
40. Mjaaland BB (2010) The plateau: imitation attack resistance of gait biometrics. In: de Leeuw E, Fischer-Hübner S, Fritsch L (eds) Policies and research in identity management. Springer, Berlin/Heidelberg, pp 100–112
41. Mjaaland BB, Bours P, Gligoroski D (2011) Walk the walk: attacking gait biometrics by imitation. In: Burmester M, Tsudik G, Magliveras S, Ilić I (eds) Information security. Springer, Berlin/Heidelberg, pp 361–380
42. Monaco JV, Ali ML, Tappert CC (2015) Spoofing key-press latencies with a generative keystroke dynamics model. In: 2015 IEEE 7th international conference on biometrics theory, applications and systems (BTAS), pp 1–8. https://doi.org/10.1109/BTAS.2015.7358795
43. Muaaz M, Mayrhofer R (2017) Smartphone-based gait recognition: from authentication to imitation. IEEE Trans Mobile Comput 16(11):3209–3221. https://doi.org/10.1109/TMC.2017.2686855
44. Neal TJ (2017) Mobile device usage data as behavioral biometrics. http://digital-library.theiet.org/content/books/10.1049/pbse003e_ch7. https://doi.org/10.1049/PBSE003E_ch7
45. Neal TJ, Woodard DL (2017) Spoofing analysis of mobile device data as behavioral biometric modalities. In: 2017 IEEE international joint conference on biometrics (IJCB), pp 62–70. https://doi.org/10.1109/BTAS.2017.8272683
46. Neal TJ, Woodard DL (2017) Using associative classification to authenticate mobile device users. In: 2017 IEEE international joint conference on biometrics (IJCB), pp 71–79. https://doi.org/10.1109/BTAS.2017.8272684
47. Ngo TT, Makihara Y, Nagahara H, Mukaigawa Y, Yagi Y (2014) The largest inertial sensor-based gait database and performance evaluation of gait-based personal authentication. Pattern Recogn 47(1):228–237. https://doi.org/10.1016/j.patcog.2013.06.028
48. Nixon M (1999) Automatic gait recognition. In: IET conference proceedings, pp 3–3(1)
49. O'Gorman L (2003) Comparing passwords, tokens, and biometrics for user authentication. Proc IEEE 91(12):2021–2040. https://doi.org/10.1109/JPROC.2003.819611
50. Oulasvirta A, Rattenbury T, Ma L, Raita E (2012) Habits make smartphone use more pervasive. Pers Ubiquit Comput 16(1):105–114. https://doi.org/10.1007/s00779-011-0412-2

51. Paul G, Irvine J (2015) Fingerprint authentication is here but are we ready for what it brings? IEEE Consumer Electronics Magazine
52. Rahman K, Balagani KS, Phoha VV (2011) Making impostor pass rates meaningless: a case of snoop-forge-replay attack on continuous cyber-behavioral verification with keystrokes. In: CVPR 2011 workshops, pp 31–38 https://doi.org/10.1109/CVPRW.2011.5981729
53. Rahman KA, Balagani KS, Phoha VV (2013) Snoop-forge-replay attacks on continuous verification with keystrokes. IEEE Trans Inf Forensics Secur 8(3):528–541. https://doi.org/10.1109/TIFS.2013.2244091
54. Ratha NK, Connell JH, Bolle RM (2001) Enhancing security and privacy in biometrics-based authentication systems. IBM Syst J 40(3):614–634. https://doi.org/10.1147/sj.403.0614
55. Ren Y, Chen Y, Chuah MC, Yang J (2013) Smartphone based user verification leveraging gait recognition for mobile healthcare systems. In: 2013 IEEE international conference on sensing, communications and networking (SECON), pp 149–157. https://doi.org/10.1109/SAHCN.2013.6644973
56. Revathi A, Jeyalakshmi C, Thenmozhi K (2018) Person authentication using speech as a biometric against play back attacks. Multimed Tools Appl. https://doi.org/10.1007/s11042-018-6258-0
57. Samangouei P, Patel VM, Chellappa R (2017) Facial attributes for active authentication on mobile devices. Image Vis Comput 58:181–192. https://doi.org/10.1016/j.imavis.2016.05.004
58. Serwadda A, Phoha VV (2013) Examining a large keystroke biometrics dataset for statistical-attack openings. ACM Trans Inf Syst Secur 16(2):8:1–8:30. https://doi.org/10.1145/2516960
59. Serwadda A, Phoha VV, Kiremire A (2011) Using global knowledge of users' typing traits to attack keystroke biometrics templates. In: Proceedings of the thirteenth ACM multimedia workshop on multimedia and security, Sec'11. ACM, New York, pp 51–60. https://doi.org/10.1145/2037252.2037263
60. Shi E, Niu Y, Jakobsson M, Chow R (2011) Implicit authentication through learning user behavior. In: Information security, vol 6531. Springer, pp 99–113. https://doi.org/10.1007/978-3-642-18178-8_9
61. Shin KI, Park JS, Lee JY, Park JH (2012) Design and implementation of improved authentication system for android smartphone users. In: 2012 26th international conference on advanced information networking and applications workshops (WAINA), pp 704–707. https://doi.org/10.1109/WAINA.2012.31
62. Sitová Z, Šeděnka J, Yang Q, Peng G, Zhou G, Gasti P, Balagani KS (2016) HMOG: new behavioral biometric features for continuous authentication of smartphone users. IEEE Trans Inf Forensics Secur 11(5):877–892. https://doi.org/10.1109/TIFS.2015.2506542
63. Stefan D, Shu X, Yao DD (2012) Robustness of keystroke-dynamics based biometrics against synthetic forgeries. Comput Secur 31(1):109–121. https://doi.org/10.1016/j.cose.2011.10.001
64. Stokkenes M, Ramachandra R, Busch C (2016) Biometric authentication protocols on smartphones: an overview. In: Proceedings of the 9th international conference on security of information and networks, SIN'16. ACM, New York, pp 136–140
65. Tari F, Ozok AA, Holden SH (2006) A comparison of perceived and real shoulder-surfing risks between alphanumeric and graphical passwords. In: Proceedings of the second symposium on usable privacy and security, SOUPS'06. ACM, New York, pp 56–66. https://doi.org/10.1145/1143120.1143128
66. Tilley A (2016) Samsung goes beyond the fingerprint with an iris scanner in the note 7. http://www.forbes.com/sites/aarontilley/2016/08/02/samsungs-note-7-iris-scanner/#69a119eb7355
67. Venugopalan S, Savvides M (2011) How to generate spoofed irises from an iris code template. IEEE Trans Inf Forensics Secur 6(2):385–395. https://doi.org/10.1109/TIFS.2011.2108288
68. Wu JS, Lin WC, Lin CT, Wei TE (2015) Smartphone continuous authentication based on keystroke and gesture profiling. In: Proceedings of international security technology (ICCST) Carnahan conference, pp 191–197. https://doi.org/10.1109/CCST.2015.7389681
69. Wu Z, Evans N, Kinnunen T, Yamagishi J, Alegre F, Li H (2015) Spoofing and countermeasures for speaker verification: a survey. Speech Commun 66:130–153. https://doi.org/10.1016/j.specom.2014.10.005

70. Xu Y, Lin M, Lu H, Cardone G, Lane N, Chen Z, Campbell A, Choudhury T (2013) Preference, context and communities: a multi-faceted approach to predicting smartphone app usage patterns. In: Proceedings of the 2013 international symposium on wearable computers, ISWC'13. ACM, New York, pp 69–76. https://doi.org/10.1145/2493988.2494333
71. Yang Y, Clark GD, Lindqvist J, Oulasvirta A (2016) Free-form gesture authentication in the wild. In: Proceedings of the 2016 CHI conference on human factors in computing systems, CHI'16. ACM, New York, pp 3722–3735. https://doi.org/10.1145/2858036.2858270
72. Yazji S, Scheuermann P, Dick RP, Trajcevski G, Jin R (2014) Efficient location aware intrusion detection to protect mobile devices. Pers Ubiquit Comput 18(1):143–162. https://doi.org/10.1007/s00779-012-0628-9
73. Yu Z, Olade I, Liang HN, Fleming C (2016) Usable authentication mechanisms for mobile devices: an exploration of 3D graphical passwords. In: Proceedings of international conference on platform technology and service (PlatCon), pp 1–3. https://doi.org/10.1109/PlatCon.2016.7456837
74. Zhang D (2014) Trustfa: trustzone-assisted facial authentication on smartphone. Technical report Dec 2014
75. Zhang Z, Hu M, Wang Y (2011) A survey of advances in biometric gait recognition. In: Sun Z, Lai J, Chen X, Tan T (eds) Biometric recognition. Springer, Berlin/Heidelberg, pp 150–158
76. Zhao S, Zhang Q, Hu G, Qin Y, Feng D (2014) Providing root of trust for arm trustzone using on-chip SRAM. In: Proceedings of the 4th international workshop on trustworthy embedded devices, TrustED'14. ACM, New York, pp 25–36. https://doi.org/10.1145/2666141.2666145
77. Zimmermann V, Gerber N (2017) "if it wasn't secure, they would not use it in the movies" – security perceptions and user acceptance of authentication technologies. In: Tryfonas T (ed) Human aspects of information security, privacy and trust. Springer International Publishing, Cham, pp 265–283

Personalized Data Minimization Assurance Using Bluetooth Low Energy

Evangelos Sakkopoulos, Zafeiria-Marina Ioannou, and Emmanouil Viennas

Abstract Mobile identity applications allow people to use a mobile phone as a form of secure digital identity (ID) card for identification purposes. In this paper, we present a novel transferring method for identity data such as electronic passport or other identification document data between two mobile devices, i.e. mobile identity holder and reader, over a BLE channel and propose the definition of a new GATT (Generic Attributes) profile suitable for mobile identity applications. Using the proposed approach, we show that BLE standard profiles can simplify and speed up mobile identity data exchange for several use cases.

1 Introduction

Mobile identity applications allow people to use a mobile device as a form of secure digital identity (ID) card. Like the identity card, the mobile identity application can be used for the identity verification of the holder by a reader device.

BLE is a light-weight subset of classic Bluetooth and was introduced by the Bluetooth Special Interest Group (SIG) [1] as part of the Bluetooth core specification. BLE has gained very high momentum, as witnessed by its widespread presence in smartphones, wearables and other consumer electronics devices. Nowadays, most of the mobile operating systems including iOS, Android, as well as macOS, Linux, Windows, natively support BLE.

In this paper, we present an extended version of a proposed transferring method for identity data [2–4]. Identity data for face to face encounters are usually stored in electronic document as ePassports, eIDAS documents [5, 6]. Only recently the idea to store personal data is including mobile apps, i.e. a mobile identity holder

E. Sakkopoulos (✉)
Department of Informatics, University of Piraeus, Piraeus, Greece
e-mail: sakkopul@unipi.gr

Z.-M. Ioannou · E. Viennas
Computer Engineering and Informatics, Department, University of Patras, Patras, Greece

T. Bourlai et al. (eds.), *Securing Social Identity in Mobile Platforms*, Advanced Sciences and Technologies for Security Applications,
https://doi.org/10.1007/978-3-030-39489-9_3

and a mobile identity verifier/reader app. Carrying personal information into a mobile device is generally referred to as mobile ID, or mobile Passport. It is a work under research and development to finalize a common cross-world approach on how to transform personal information typically stored in a travel identification document into a mobile identification app that will be possibly read by any third party.

In order to facilitate identity information transfer, we have proposed that existing architectures can be helpful such as the GATT profile specification for BLE. GATT profiles are available for a number of predefined services and use cases such as subscribing to a headset or transferring data from a wristband. GATT profiles are data specific and also incorporate certain orchestration for each case. In our work [7] we have introduced a new GATT profile that may transfer personal information data over a BLE channel. We have proposed in [7] the definition of a new GATT (Generic Attributes) profile suitable for mobile identity applications. In this work, we provide further details on data minimization based on that approach. Data minimization is a core topic in privacy protection processes especially as governed by GDPR the EU Regulation on General Data Protection. The proposed method simplifies and speeds up the process of identity data transferring between mobile devices.

The rest of the paper is organized as follows: in Sect. 2 we present the related work. Section 3 describes an overview of BLE protocol. Sections 4 and 5 presents the mobile ID GATT profile definition and a proposed mobile identity data transferring method over BLE. In Sect. 5 provides practical examples of use cases for the proposed method using real data; it includes also a data minimization discussion that shows the potential of our approach in enforcing privacy protection. Finally, Sect. 6 presents our conclusions and thoughts for future work.

2 Related Work

Electronic Machine Readable Travel Documents (eMRTDs), also called as ePassports or biometric passports, differ from the ordinary passports as they additionally has an embedded contactless Integrated Circuit (IC). The IC stores the personal data printed on the passport data page, one or more biometric features of the passport holder e.g. facial image, fingerprints, iris and a security object, a digitally signed file to check authenticity and integrity of content. The data encoded in the IC are protected by Public Key Infrastructure (PKI) cryptographic technology against tampering and unauthorized reading and cloning by security mechanisms.

The International Civil Aviation Organization (ICAO) has specified the Logical Data Structure (LDS) in the part 10 of Doc 9303 [8] for eMRTDs that defines a standardized data structure for the organization of data stored in the contactless IC, including identity and biometric data. This was to achieve the global interoperability for electronic reading of the eMRTDs. The data stored in LDS are organized as a collection of Data Groups (DG) each one of which consists of standardized Data Elements and is stored in a separate Elementary File (EF). More specifically, the

Table 1 Logical data structure – data groups

Data group	Description
DG1	Details recorded in MRZ
DG2	Encoded face
DG3	Encoded fingers
DG4	Encoded eyes
DG5	Displayed portrait
DG6	Reserved for future use
DG7	Displayed signature or usual mark
DG8	Data features
DG9	Structure features
DG10	Substance features
DG11	Additional personal details
DG12	Additional document details
DG13	Optional details
DG14	Security options
DG15	Active authentication public key info
DG16	Persons to notify
EF.COM	Common data
EF.SOD	Document security object

LDS defines 16 Data Groups, DG1-DG16, where the two first are mandatory while all the other are optional. Additionally, there are two more mandatory Elementary Files, EF.COM and EF.SOD (Table 1).

For instance, DG1 defines the details of the Machine Readable Zone (MRZ), DG2 contains the encoded face image of the passport holder, EF.COM consists of the version information and a list of the existing Data Groups and EF.SOD, which is used to validate the integrity of stored data, contains the hashes of existing data groups as well as a digital signature of the hashes. Data groups 3 and 4 are optional and contain additional biometric features, i.e. the encoded fingerprints and encoded irises of the passport holder.

Currently available solutions for reading eMRTDs or other electronic identity documents using mobile devices are based on Near-field communication (NFC) [9, 10] as a transmission protocol. NFC is a short-range wireless connectivity technology and its standards are provided by the NFC forum.

An NFC enabled device can work in three operation modes:

- *reader/writer*: an NFC-enabled device is capable of reading information stored on NFC tags embedded e.g. in labels or smart posters.
- *peer-to-peer*: Two NFC-enabled devices can communicate directly with each other and exchange data.
- *card emulation*: an NFC-enabled device can act like a contactless smart card. This mode enables contactless payments and ticketing.

NFC relies on the ISO/IEC 14443 standard that defines proximity identification cards and the transmission protocols for communicating with it. Also, NFC enabled devices are compatible with ISO/IEC 14443. Consequently, eMRTDs are compatible with NFC enabled devices that support the reader/writer operation mode. Additionally, an eMRTD can be implemented on an NFC enabled mobile device using the card emulation operation mode.

The communication between mobile reader devices and contactless smart cards is achieved through commands and responses known as Application Protocol Data Units (APDUs) in order to establish a secure messaging mechanism or to receive each one of the data group files of the LDS. The basic structure of an APDU is defined by ISO/IEC 7816-4. There are two categories of APDUs, command APDUs and response APDUs (Tables 2 and 3).

The Regulation (EU) N°910/2014 on electronic identification and trust services for electronic transactions in the internal market (eIDAS Regulation [11]) adopted on 23 July 2014 provides a predictable regulatory environment to enable secure and seamless electronic interactions between businesses, citizens and public authorities.

Table 2 Command APDU structure

Field name	Length (bytes)	Description
Header		
CLA	1	Class of instruction: indicates the type of command, e.g. interindustry or proprietary
INS	1	Instruction code: indicates the specific command, e.g. read binary
P1-P2	2	Instruction parameters 1 and 2 for the command, e.g. offset into file at which to write the data
Body		
Lc	0, 1 or 3	Encodes the number (Nc) of bytes present in the data field of the command:
		0 bytes denotes Nc = 0
		1 byte with a value from 1 to 255 denotes Nc with the same value
		3 bytes, the first of which must be 0, denotes Nc in the range 1 to 65 535 (all three bytes may not be zero)
Data field	Nc	N_c String of bytes sent in the data field of the command
Le	0–3	Maximum number of bytes (Ne) expected in the data field of the response to the command
		0 bytes denotes Ne = 0
		1 byte in the range 1 to 255 denotes that value of Ne, or 0 denotes Ne = 256
		2 bytes (if extended Lc was present in the command) in the range 1 to 65 535 denotes Ne of that value, or two zero bytes denotes 65 536
		3 bytes (if Lc was not present in the command), the first of which must be 0, denote Ne in the same way as two-byte Le

Table 3 Response APDU structure

Field name	Length (bytes)	Description
Response data	Nr (at most Ne)	Response data
SW1-SW2 (Response trailer)	2	Status Bytes 1 and 2: command processing status and command processing qualifier, e.g. the value 90 00 (hexadecimal) indicates success

In this regard, the eIDAS Regulation:

- ensures that people and businesses can use their own national electronic identification schemes (eIDs) to access public services in other EU countries where eIDs are available.
- creates a European internal market for electronic trust services – namely electronic signatures, electronic seals, time stamp, electronic delivery service and website authentication – by ensuring that they will work across borders and have the same legal status as traditional paper based processes.

However, no systematic and specific provisions [12] have been made in order to transform in any way eID into mobile ID (mID) before 2019 especially by reusing BLE standard architecture.

In this work, we propose a new data transferring method for mobile identity applications based on the Bluetooth Low Energy protocol. The method based on a new defined GATT profile simplifies and speeds up the process of data transferring between two mobile devices, the mobile ID holder and mobile ID reader. The key advantages of our approach is that it reduces the number of commands and responses that are required be sent in order to the identity data to be transferred when it is compared to typical chip based data exchange using APDU commands.

3 Overview of BLE Protocol

In this section we present the key architectural approach for the BLE Protocol. Instead of using the BLE transmission protocol in an agnostic manner i.e. as a transport medium, BLE specs set the basis for our proposed data for identification mobile exchange. The BLE protocol stack is composed of three main layers: Controller, Host and Application, as depicted in Fig. 1.

In particular, the Controller includes Physical Layer (PHY) that controls radio communication of transmitting/receiving data and Link Layer (LL) that defines packet structure, includes the state machine and radio control, and provides link layer-level encryption. The Host Controller Interface (HCI) provides a standard interface for communication between the Controller and Host layers. The Host consists of the following layers: Logical Link Control and Adaptation Protocol (L2CAP), Security Manager (SM), Attribute Protocol (ATT), Generic Attribute Pro-

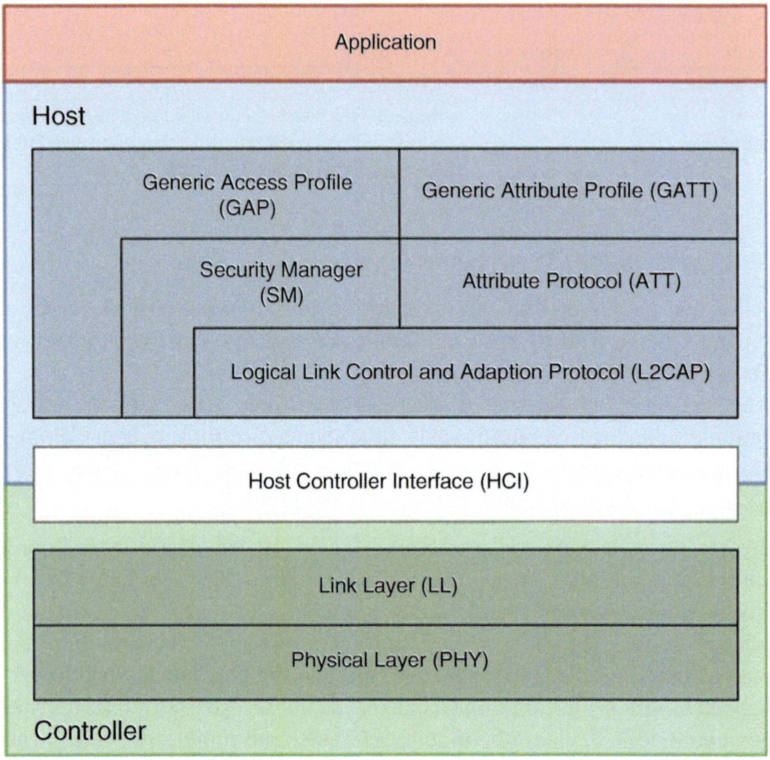

Fig. 1 Bluetooth low energy (BLE) protocol stack

file (GATT) and Generic Access Profile (GAP). The L2CAP supports higher level protocol multiplexing, performs packet segmentation and reassembly operations, along with the conveying of quality of service information. The SM defines methods and protocols for device pairing and key distribution. The ATT defines the protocol of transferring the attribute data and provides GATT related procedures such as read, write and notification. The GATT built on the top of the ATT defines and creates the types of attributes and how they can be used for a given application while ATT provides an information exchange mechanism between devices in the form of attributes. The GAP is responsible for the advertisement and connection functionality, allows a device to be visible to other devices and manages the communication between devices.

Also, according to GAP a device can operate in one or more of the following roles specified in GAP:

Broadcaster: A device that broadcasts advertising data packets.

Observer: A device that listens to BLE devices and processes data from the advertising packets send by broadcaster. There is no connection between a broadcaster and observer.

Peripheral: A device that advertises its presence so central devices can establish a connection. After connecting, peripherals no longer broadcast data to other central devices and stay connected to the device that accepted connection request.
Central: A device that initiates a connection with a peripheral device by first listening to the advertising packets. A central device can connect to many other peripheral devices.
Finally, the Application layer works as the interface between the user application and BLE protocol stack.

Our focus is on GATT layer since the goal is to extend original BLE approach widely used for typically smaller amounts of information exchange with a new profile that has appropriate format and placeholders for all needed identification information. A profile defines a specific use case, roles and general behaviors based on the GATT functionality. The GATT defines a hierarchical data structure that is exposed to connected BLE devices. A GATT profile consists of two main elements: service and characteristic. The service is a collection of one or more characteristics and characteristics consist of data value, a set of properties which defines the operations that characteristic supports as well as a set of permissions regarding the security. The available properties of a characteristic are:

Read: if set, allows reads of characteristic value.
Write: if set, allows writes of characteristic value (with or without response).
Notify: if set, allows notifications of a characteristic value (without acknowledgement) when the characteristic value has been updated.
Indicate: if set, allows indications of a characteristic value (with acknowledgement) when the characteristic value has been updated.

A characteristic may also contain one or more descriptors which give information about the characteristic or allow the configuration of a behavior involving the characteristic. For example, notifications or indications can be enabled or disabled by using a descriptor called the Client Characteristic Configuration Descriptor. Generally, the profile is a group of services and services contain characteristics where each characteristic contains values, properties and additional description. The hierarchical data structure of a GATT profile is shown in Fig. 2.

Similar to GAP, there are two GATT roles: GATT Server and GATT Client. GATT Server is a device that stores attributes and makes them available when a device in GATT Client role sends a request. GAP and GATT roles are essentially independent of one another.

There are the following methods for transferring data between a GATT Server and a GATT Client over a BLE connection:

(a) *Write*: The Client sends data to the Server by writing the value to a characteristic,
(b) *Read*: The Server sends data to the Client when the Client sends a command to read the value of a characteristic,
(c) *Notifications/Indications*: The Client receives data via notifications or indications when it has enabled these operations. In this case, the Client receives data without the need of sending a request.

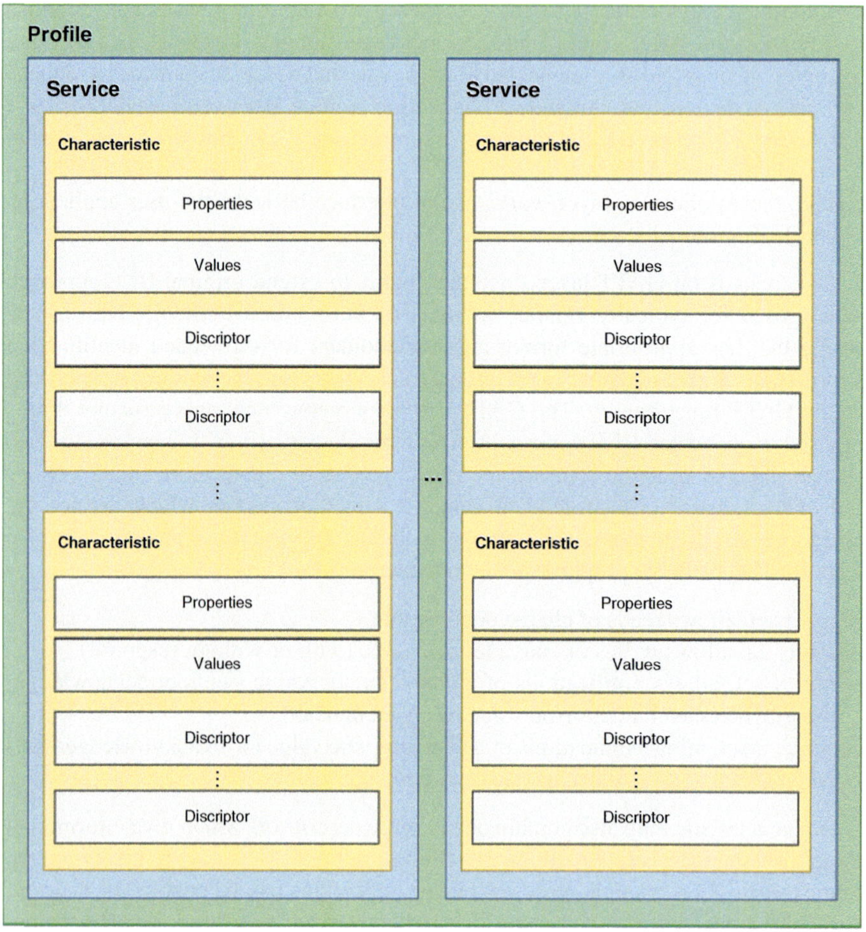

Fig. 2 GATT profile hierarchy

The Bluetooth specification provides several profiles but also allows developers to define their own profiles for use cases that are not covered by the existing SIG-defined profiles. In the following section, we present the definition of a custom profile suitable for mobile identity applications.

4 Mobile Identification Data Profile

The proposed Profile is used to transmit mobile identification data over a BLE link. One service needs to be defined over the GATT layer with two characteristics.

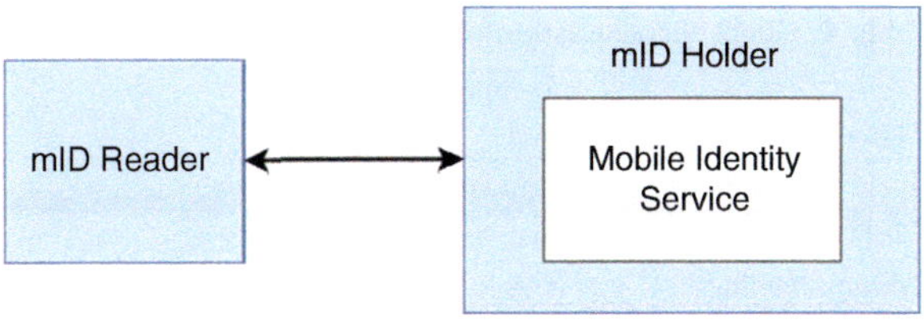

Fig. 3 Relationship between service and profile roles

The profile defines two roles: the reader and the data owner. The reader is the device that sends commands to the data owner requesting retrieval. The data owner receives the commands and sends responses via notifications to the reader device. The data owner includes the Mobile Identity Service.

In our server based approach, the owner shall be a GATT Server while the reader shall be a GATT Client. The specification allows the positioning of client and server to be vice versa but that does not affect the validity and effectiveness of our approach. Figure 3 shows the relationships between service and two profile roles.

The data holder device instantiates one and only one Mobile Identity Service.

So the data holder shall use the GAP Peripheral role, while the reader shall use the GAP Central role.

4.1 Mobile Identification Data Service

The mobile identification data – mID Service enables to send data to the Reader over a BLE link. The service is not dependent upon any other services. Requirements include mandatory server objects for Write Characteristic Value, Notifications, Read Characteristic Descriptors, Write Characteristic Descriptors.

4.2 Service Characteristics

The Service contains the following two characteristics:

1. **Mobile Identity Data**

This characteristic is used to send mobile identity data to the mID Reader Device.

It includes mandatory properties such as Notify and Mandatory Descriptors such as Client Characteristic Configuration Requirement with permissions read and write.

In order to facilitate the personal information data transfer in an organized manner we devised the following structure for the value fields (Table 4):

Table 4 Client characteristic configuration permissions

Name	Accept to send mID data		Deny sending mID data
Field requirement	Mandatory		Mandatory
Format	Byte array		uint16
Minimum value	N/A		N/A
Maximum value	N/A		N/A
Additional information	Header (1 byte)	Data	Header (1 byte)
	1 h	Number of blocks to be sent (4 bytes)	3 h
		Length of mID data to be sent (4 bytes)	
	2 h	One block of mID data	

Table 5 Mobile identity control point value fields

Name	Request	
Field requirement	Mandatory	
Format	uint48	
Minimum value	N/A	
Maximum value	N/A	
Additional information	Bit	Description
	0–1	Mode: 00 for offline or 01 for online
	2–3	Type of request: 00 for requesting complete mID data or 01 for requesting age verification data
	4–35	Reader ID number
	36–47	Reserved for future use

The reader shall control the configuration of the notifications via the Client Characteristic Configuration descriptor of the Mobile Identity Data characteristic and shall be able to receive multiple notifications from the data owner.

The Reader shall determine the data content of the Mobile Identity Data characteristic based on the header of value field. In particular, a header with value:

(a) *0x01*: indicates that mID Reader has accepted the request for sending mID data and the receiving data packet contains the number of blocks as well as the length of mID data to be received. These two values should be used by mID Reader to verify the correct reception of received data.
(b) *0x02*: indicates that the receiving data packet contains a block of mID data.
(c) *0x03*: indicates that the mID Holder has rejected the request to send mID data.

2. **Mobile Identity Control Point**

The mID Control Point Characteristic allows the mID Reader to send request commands to the mID Holder Device for receiving mobile identity data. It needs write permissions and the value fields is proposed to follow the following structure (Table 5).

The reader sends the request to receive data by writing a value to the Control Point characteristic. The value may contain additional information, such as the operation mode of the Reader device (e.g. online/offline or other supported operation modes), the type of request, for example requesting complete data or only age verification data including age attestation/date of birth and facial image, as well as the reader ID Number which could be used by data owner device for verifying the identity of the device that makes the request.

5 Data Transferring Method

The reader device in the Central role scans, looking for advertisement, while the data owner device in the Peripheral role makes the advertisement of a Service and its characteristics. Following we describe the method step by step in order to achieve successful data exchange.

(a) Service Discovery

The mID Reader device performs service discovery using the GATT Discover All Primary Services sub-procedure or the GATT Discover Primary Service by Service UUID sub- procedure to discover the Mobile Identity Service with "mID Service" for the service UUID.

(b) Characteristic Discovery

The mID Reader device performs the GATT Discover All Characteristics of a Service sub-procedure or the GATT Discover Characteristics by UUID sub-procedure to discover all the characteristics and characteristic descriptors of the mID Service.

(c) Sending the request command

When a BLE connection has been established between the mID Holder and mID Reader devices, the mID Reader transmits a request command to the mID device to receive mobile identity data. Before sending any command, the mID Reader shall set the Client Characteristic Configuration Descriptor of the Mobile Identity Data Characteristic to enable Notifications. It is also recommended that the mID Reader (Client) should request an increase of the Maximum Transmission Unit (MTU) to the maximum possible value.

The MTU corresponds to the maximum size of a single packet that can be transmitted over the BLE connection and the default value of MTU is 23 bytes. During this procedure, the Client informs the Server about its maximum supported receive MTU size and the Server responses with its maximum supported receive MTU size. The procedure can only be initiated by the Client and must be performed on each BLE connection. After the exchange procedure, the MTU value for the current connection is set as the minimum value of the Client MTU and Server MTU values. For example, if the Client maximum supported MTU size is 250 bytes and

Server maximum supported MTU size is 200 bytes, then the MTU size has to be set to 200. Increasing the MTU size increases the size of data that can be transmitted in a single packet and further speeds up the data transferring process. The increase of MTU size is an important step when transferring large amounts of data, as in the case of mobile identity data.

For sending a request to the mID device, the mID Reader writes a value to the Mobile Identity Control Point Characteristic including the operation mode (online/offline) based on internet connection availability as well as the reader identification number.

(d) Receiving the request and sending mID Data

The mID device receives the incoming request from the mID Reader and the mID Holder must accept or deny the request for sending mID data. The mID data has to be in the form of a byte array. The mID device reads the value of the mID Control Point characteristic including the operation mode and reader identity number.

In case of acceptance the mID device:

1. The length of mID Data as well as the number of blocks to be sent according to the MTU size of the connection and then transferred to mID Reader by writing values on the Mobile Identity Data Characteristic and sending a notification. In particular, the number of blocks to be send can be calculated using the following Eq. (1):

$$\text{Number of Blocks} = \text{Length of mID Data} / (\text{MTU}-3-1) \quad (1)$$

where 3 is the number of bytes required as header when sending a Write, Read, Notification or Indication packets, 1-byte for OP-Code indicating the ATT Operation and 2-bytes for Attribute Handle for identification of the data and 1 is the header defined for mID data.

2. The mID Data are fragmented into blocks according to the MTU size and each block is transmitted by writing the value on the Mobile Identity Data Characteristic and sending a notification. Between writing blocks on the characteristic and sending notifications, a delay of about 100ms must be added.

In case of denial, a byte array with value 3 h is sent by writing the value on the Mobile Identity Data Characteristic and sending a notification.

(e) Receiving mID Data

The mID Reader device receives via notifications the denial response or the length of mID Data, the number of blocks as well as all the blocks of mID Data. At the end of the transmission process, the data blocks have to be merged into one and according to the length of mID Data as well as number of blocks values the correctness of the mID data has to be verified. The mID Reader is able to request a retransmission of data if the received data are not correct. Finally, the mID Reader closes the BLE connection between two devices. Figure 4 shows a sequence diagram of the described mID data transferring method.

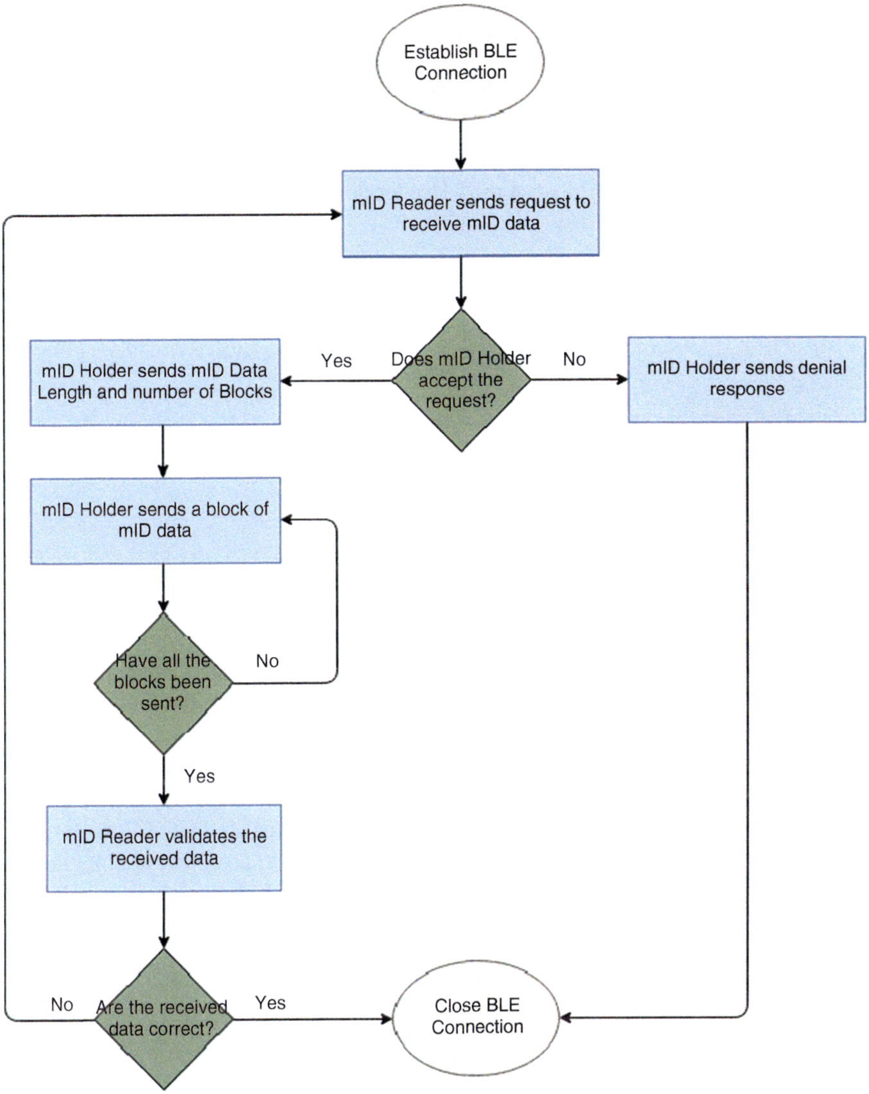

Fig. 4 mID data transferring method

6 Use Case Example

In this section, we provide practical examples of use cases presenting the data transferring method with real data.

6.1 *Complete mID Data Request*

Suppose that the complete mobile identity data is required by the mID Reader and that the mID data including full name, date of birth, date of issue, date of expiry, document id, address, facial image, gender, eye color, height and weight are represented in JSON format on mID Holder device as following:


```
{"LASTNAME":"SMITH",
"FIRSTNAME":"JANE",
"DATEOFBIRTH":"1986-03-30",
"DATEOFISSUE":"2008-06-19",
"DATEOFEXPIRY":"2051-03-30",
"DOCUMENTID":"111111111",
"ADDRESS":"1600 Pennsylvania Ave NW, Washington DC 20005",
"FACIALIMAGE":"...",
"GENDER":"FEMALE",
"EYES":"BLUE",
"HEIGHT":65,
"WEIGHT":111}
```


and are stored in a byte array of size 8000. The mID Reader device can operate in two different modes, online and offline, according to the internet connection availability. In offline mode, all the fields of mID data must be sent by the mID holder, while in the online mode only the fields of document id, date of birth and date of expiry in JSON format must be sent as the mID Reader can then retrieve all the mID dataset from the server providing the three fields as parameter.

After a BLE connection has been successfully established between mID Holder and mID Reader device, the mID Reader enables notifications using the Client Characteristic Configuration Descriptor of the Mobile Identity Data Characteristic and further requests an increase of the MTU size. Suppose, also, that the maximum supported MTU size between the two devices has set to 512 bytes.

6.1.1 Offline Mode

The mID Reader sends a request command to the mID device to receive mID data by writing the following value to the Mobile Identity Control Point Characteristic (Table 6).

The first two bits of the value indicate that the mID Reader device works in the offline mode, the following two bits determine the requested type of data i.e. the

Table 6 Offline request command

Mode	00
Type of data	00
Reader ID number	00000111010110111100110100010101
Reserved for future use	000000000000

Table 7 First packet of data

Header	00000001
Number of blocks	00000000000000000000000000010000
Length of mID data to be sent	00000000000000000001111101000000

Table 8 Packets of data

Header	00000010
Block of mID data	(maximum block length: 512–3-1 = 508 bytes)

complete mID data, while the last 32 bits correspond to the reader identity number with the value 123456789_{10}.

The mID device receives the incoming request and reads the value sent to the mID Control Point characteristic and checks the operation mode, the type of data, as well as the validity of reader id number.

(a) In case of acceptance:

Suppose that the mID holder accepts the request. The number of blocks to be sent is calculated following the Eq. (1) as $8000/(512 - 3 - 1)$, or 16 blocks. Subsequently, the data owner sends the first packet including the header, the length of Data as well as the number of blocks to be sent by writing the following value on the Mobile Identity Data Characteristic and sending a notification (Table 7).

In the following 16 packets, the Holder device sends the blocks of data as follows (Table 8):

by writing the value on the Mobile Identity Data Characteristic and sending a notification.

(b) In case of deny:

Suppose that the mID holder denies the request e.g. due to security reasons, such as in the case of an invalid reader identity number. In this case, the mID holder device sends the denial response value by writing it on the Mobile Identity Data Characteristic and sending a notification (i.e. 00000011).

6.1.2 Online Mode

In online operation mode, the mID Reader sends the request command to the mID device to receive mID data by writing the following value to the Mobile Identity Control Point Characteristic (Table 9):

The first two bits of the value indicate that the Reader device works in the online mode.

The data owner device receives the incoming request and reads the value of the Control Point characteristic and checks the operation mode as well as the validity of reader id number. In case of acceptance, the Reader is required to send only the following three fields of data in JSON format.

Table 9 Denial response

Mode	01
Type of data	00
Reader ID number	00000111010110111100110100010101
Reserved for future use	000000000000

Table 10 First packet of mID data

Header	00000001
Number of blocks	00000000000000000000000000000001
Length of mID data to be sent	00000000000000000000000001010001


```
{"DATEOFBIRTH":"1986-03-30",
"DATEOFEXPIRY":"2051-03-30",
"DOCUMENTID":"111111111"}
```


The length of byte array is 81 and, consequently, all the data can be sent in one block.

The mID Holder device sends in the first packet the length of data and the number of blocks by writing the following value on the Mobile Identity Data Characteristic and sending a notification (Table 10).

In the second packet, the mID data are sent in one block by writing the following value i.e. 00000010 on the Mobile Identity Data Characteristic and sending a notification (81 bytes).

Finally, the mID Reader device can retrieve all the mID dataset from the server providing the received fields as parameters.

6.2 Age Verification Data Request

Suppose that the age verification data is required by the mID Reader to verify the age of the customer for example in case of the purchase of certain commodities such as alcohol or tobacco products. The age verification data including date of birth and facial image are represented in JSON format on mID Holder device as following:


```
{"DATEOFBIRTH":"1986-03-30",
"FACIALIMAGE":"..."}
```


and are stored in a byte array of size 5000.

After a BLE connection has been successfully established between mID Holder and mID Reader device and the increase of the MTU size, the mID Reader sends a request command to the mID device to receive the age verification data by writing the following value to the Mobile Identity Control Point Characteristic (Table 11).

The first two bits of the value indicate that the mID Reader device works in the offline mode, the following two bits determine the requested type of data i.e. the

Table 11 Age verification data request command

Mode	00
Type of data	01
Reader ID number	00000111010110111100110100010101
Reserved for future use	000000000000

Table 12 First packet of mID data

Header	00000001
Number of blocks	00000000000000000000000000001010
Length of mID data to be sent	00000000000000000001001110001000

age verification data, while the last 32 bits correspond to the reader identity number with the value 123456789_{10}.

The mID device receives the incoming request and reads the value sent to the mID Control Point characteristic and checks the operation mode, the requested type of data, as well as the validity of reader id number.

(a) In case of acceptance:

Suppose that the mID holder accepts the request. The number of blocks to be sent is calculated following the Eq. (1) as 5000/ (512 – 3 – 1), or 10 blocks. Subsequently, the mID sends the first packet including the header, the length of mID Data as well as the number of blocks to be sent by writing the following value on the Mobile Identity Data Characteristic and sending a notification (Table 12).

In the following 10 packets, the mID Holder device sends the blocks of data using header 00000010 and maximum 508 bytes by writing the value on the Mobile Identity Data Characteristic and sending a notification.

(b) In case of deny:

Suppose that the mID holder denies the request e.g. due to security reasons, such as in the case of an invalid reader identity number. In this case, the mID holder device sends the denial response value by writing it on the Mobile Identity Data Characteristic and sending a notification.

7 Conclusion

In this paper, we present a new method for transferring mobile identity data using the architectural approach that governs Bluetooth Low Energy specifications and not merely using BLE as a link or transport medium. In this way we enable advantages that come with the capabilities that BLE GATT profiles provide for security, and integrity. Further, we minimize the number of handshake steps needed in comparison to a chip based data transporting. The proposed method is using a new defined GATT profile specially designed for identification data which are by design larger than typical vital sign or other BLE devices logged data. As a result, it simplifies and further speeds up the process of transferring mobile identity data between two mobile devices by reducing the number of commands and responses that are required be sent in order to the identity data to be transferred. Our future work focusses in the following two directions: (a) standardization of the proposed GATT profile as well as the data transferring solution, and (b) extension of the method in order to include further capabilities, such as commands.

References

1. Bluetooth.https://www.bluetooth.com/
2. Krimpe J (2014) Mobile ID: crucial element of m-government. In: Proceedings of the 2014 conference on Electronic Governance and Open Society: challenges in Eurasia (EGOSE'14). ACM, New York, pp 187–194
3. Benes F, Stasa P, Svub J, Unucka J, Rhee J, Vojtech L (2017) Object localization for determining Customer's behavior: auto-ID based approach. In: Proceedings of the 14th EAI international conference on mobile and ubiquitous systems: computing, networking and services (MobiQuitous 2017). ACM, New York, pp 468–477
4. Průša J (2015) E-identity: basic building block of e-government. In: 2015 IST-Africa conference, Lilongwe, pp 1–10
5. Bieker F, Hansen M (2014) Privacy-preserving authentication solutions – best practices for implementation and EU regulatory perspectives. In: eChallenges e-2014 conference proceedings, Belfast, pp 1–10
6. Berbecaru D, Lioy A (2018) On integration of academic attributes in the eIDAS infrastructure to support cross-border services. In: 2018 22nd International Conference on System Theory, Control and Computing (ICSTCC), Sinaia, Romania, pp 691–696
7. Sakkopoulos E, Ioannou Z-M, Viennas E (2018) Mobile personal information exchange over BLE. In: The 9th international conference on information, intelligence, systems and applications, 23–25 July, Zakynthos, Greece
8. International Civil Organization (2015) Part 10: logical data structure for storage of biometrics and other data in the contactless integrated circuit (IC). In: Doc 9303, machine readable travel documents, 7th edn. International Civil Organization, Montréal
9. NFC – ISO/IEC 14443-1:2016 Identification cards – contactless integrated circuit cards – proximity cards. https://www.iso.org/standard/73599.html
10. NFC Forum. https://nfc-forum.org/
11. eIDAS. https://ec.europa.eu/futurium/en/content/eidas-regulation-regulation-eu-ndeg9102014
12. Gunawong P, Gao P (2010) Challenges of egovernment in developing countries: actor-network analysis of Thailand's smart ID card project. In: Proceedings of the 4th ACM/IEEE International Conference on Information and Communication Technologies and Development (ICTD'10). ACM, New York, NY, USA, Article 17

Part II
Mobile-Based Biometric Technologies

On Designing a Forensic Toolkit for Rapid Detection of Factors that Impact Face Recognition Performance When Processing Large Scale Face Datasets

J. Rose and T. Bourlai

Abstract Due to the overlap between the fields of forensic investigation and biometric recognition, including face recognition, there have been several interesting applications that bridge the gap between the two sciences and better connect the associated communities. These applications have been developed with the intent to assist law enforcement officers with computer assisted and biometrics related capabilities. Thus, utilizing biometric algorithms within the forensics field can support law enforcement investigations in a wide array of applications, including fingerprint comparisons, sketch-to-photo face comparisons, and even find persons of interest via soft biometrics such as scars, marks, and tattoos. In this book chapter, we focus on facial recognition, which can help provide clues when other forensic evidence is not present or available and, most importantly, help investigators eliminate the time consuming processes of interviewing potential witnesses or manually searching through thousands of mugshots to determine a suspect's identity. To aid in this mission, we propose a software toolkit to automatically and hierarchically categorize face images with a set of binary classifiers using three different attributes, which depending on their true label/condition can affect facial recognition performance. These attributes are: based on facial photo, (1) determining whether a subject's eyes are open or closed, (2) whether the subject is wearing glasses or not, and (3) whether the facial pose of the subject is either frontal or non-frontal. Our toolkit offers batch processing and therefore can aid forensic operators with a capability to rapidly categorize large scale face datasets in terms of the aforementioned attributes, and thus, determine, which individuals have a higher chance to be identified based on their face information. The proposed

J. Rose
West Virginia University, Morgantown, WV, USA
e-mail: jrose24@mix.wvu.edu

T. Bourlai (✉)
Lane Department of Computer Science and Electrical Engineering, Multispectral Imagery Lab (MILab), West Virginia University, Morgantown, WV, USA
e-mail: Thirimachos.Bourlai@mail.wvu.edu; ThBourlai@mail.wvu.edu

T. Bourlai et al. (eds.), *Securing Social Identity in Mobile Platforms*, Advanced Sciences and Technologies for Security Applications,
https://doi.org/10.1007/978-3-030-39489-9_4

forensic toolkit will allow the operators to analyze, enhance, group, or exclude face data before being used for face matching.

1 Introduction

In this section, we provide a brief introduction on forensic biometrics. Next, related work in this area, mainly face recognition and how it is affected by the three factors we are categorizing, is highlighted. The final part of this section describes our motivations and contributions for the work we present in this chapter.

The goal of both biometric recognition and forensic science is to link biological data to an individual [1]. However, the ability to use biometric systems successfully in forensic scenarios is quite challenging. The challenges in this field, often referred to as forensic biometrics, as well as their similarities and differences are well documented in [1–8]. According to [3], biometric technology plays a role in several forensic applications: the identity management and the identity verification in the criminal justice chain, the identification of missing persons from a mass disaster, and the forensic investigation and intelligence as well as the forensic evaluation of biometric evidence in court, which together form the field of forensic biometrics. More specifically, and explained in [4], forensic biometric systems are used as sorting tools which do not make any final identification decisions. For forensic face recognition scenarios, an unknown probe face image is compared to every other face image in a gallery database. The FR system computes a similarity score for the probe with each sample in the gallery and the top-K matches are returned, often ordered from most to least similar. Then the forensic investigator performs a visual inspection of each candidate from the list to determine if any of the returned faces are a match to the unknown probe, meaning that the forensic biometric system is an external tool from the manual identification process.

1.1 Related Work

Forensic biometric systems are available for many modalities including face, sketch-to-photo-face, fingerprint, ear, forensic speaker recognition, and soft biometrics like scars, marks, and tattoos (SMT). In situations where primary biometrics like face and fingerprints are not available or sufficient, tattoos, which are often collected by law enforcement to aid in identification, are commonly used. In [8], the authors propose the Tattoo-ID automatic tattoo matching and retrieval system, which extracts SIFT keypoints and then uses a matching algorithm to measure visual similarities between the probe and gallery images before retrieving the database images with the largest similarity. It proved to be a significant improvement over using the ANSI/NIST-ITL1-2011 standard that uses defined classes to query tattoo images [9].

One of the most valuable tools for forensic biometrics is face recognition [10, 11]. Biometric face recognition can aid law enforcement in several ways, including the detection of multiple records in a database, an additional method of identification when fingerprints or other information may not be available, rapid identity checks in the field, and a lead generator for investigations. Perhaps most importantly, biometric FR systems can quickly return a list of potential suspects to forensic operators who must manually perform the final identification of a suspect, leading to improved efficiency in both time and recognition accuracy. Returning accurate candidate lists is especially important due to the inherent human error when conducting face recognition. In [12], the authors test human performance on FR candidate lists of both adults and children. Results showed very poor face matching performance, with untrained participants making over 50% identification errors and trained participants making 20% fewer errors. Often face recognition scenarios require the investigator to match low quality images captured in uncontrolled conditions against a very large database. Jain et al. discuss many of the face recognition and image retrieval challenges in forensics in [6] and [7]. Some of the major challenges in unconstrained face recognition are variations in pose, expression, occlusion, age, and image quality factors such as illumination, blurriness, and brightness. To improve face recognition performance it is important to identify which images in a database have these attributes so that they may be further analyzed or enhanced. The three factors we will focus on are (1) whether a subject's eyes are open or closed, (2) whether the subject is wearing glasses or not, and (3) whether the facial pose of the subject is either frontal or non-frontal.

- **Eyes are Open or Closed:** Detecting the eyes in face images is an important step in many automated face recognition algorithms and facial landmark localization [13]. Much like face detection, the eyes have variations in appearance due to size, pose, rotation, occlusion such as glasses, opening and closure of eyes, and illumination conditions [14]. Common factors such as closed eyes and glasses can affect different eye localization methods as observed in [15–17] and therefore, FR systems. Several studies have shown that face normalization schemes based on the centers of the eyes contribute to decreased face recognition performance if eye locations are inaccurate [18] or eyeglasses are occluding the face [19]. To overcome some of these challenges, for example, law enforcement has used image editing and enhancement techniques of probe images, such as manually replacing closed eyes with open eyes to yield additional and more accurate returns, leading to thousands of arrests [20].

  The classification of open and closed eyes has applications in various fields including driver drowsiness detection, facial expression classification, and iris recognition. Extensive research has been done in this area using various methods including feature based [21–25], motion based [26–28], and appearance-based techniques [29–31]. More recently, Ji et al. [25] detected eye state by extracting contour features that are fitted by extracting sclera border points before determining eye state using a proposed eyelid closure value. In [24], a deep residual Convolutional Neural Network (CNN) structure is trained and tested with images collected in two different environments, achieving a lower equal error rate (EER) for classification when compared to other CNN methods like AlexNet and GoogLeNet, and other non-CNN based methods. The authors in [23] combine a deep neural network and a deep CNN to construct a deep integrated neural network for characterizing useful information in the eye region using a joint optimization method and a transfer learning strategy to extract effective abstract eye features and improve classification capability in uncontrolled scenarios. Their experiments showed that the proposed method outperformed current state-of-the-art methods.

- **Wearing Glasses or Not:** Eyeglasses are the most common occurrence of facial occlusions and have a significant effect on face recognition performance. Not only do eyeglasses occlude the face, eyeglass frames can also be used to intentionally fool FR systems like the frames proposed in [32]. The quick and accurate detection and, if necessary, removal of eyeglasses can be a critical factor in forensic biometric scenarios.

  The detection and removal of eyeglasses has been thoroughly studied and methods fall into two main categories, conventional handcrafted features [33–37] and deep learning approaches [38–40]. In [33], filtered edge intensities on grayscale images are used to determine the presence of glasses before using PCA reconstruction and inpainting to extract and remove the glasses respectively. Alorf and Abbott [34] used local descriptors and support vector machines to detect eye state, mouth state, and presence of glasses to achieve state-of-the-art performance when compared to CNN methods. In [36], a method for eyeglasses detection, location, and a frame discriminant based on edge information is proposed. By finding the horizontal and vertical nose bridge, the existence of eyeglasses is determined and the location found using a bidirectional edge information projection. The authors then check the existence of frames and can measure frame width based on the location of the left and right glasses. An eyeglasses detection framework based on a shallow CNN is created in [38]. Using the pretrained GoogLeNet architecture fine-tuned for images with and without eyeglasses, the learned weights from GoogLeNet are copied to the corresponding layers in the shallow CNN and used as a feature extractor to be classified by a trained linear SVM. The shallow architecture CNN reduced detection time by almost a factor of two while retaining high detection accuracy. Wang et al. [39] propose a facial obstructions removal scheme based on an Enhanced Cycle-Consistent Generative Adversarial Network (ECGAN) for face recognition. Eyeglasses are used as facial obstructions, which are detected using a CNN. The eyeglasses are then removed using the ECGAN, improving accuracy of face recognition compared to other existing approaches.
- **Pose is Frontal or Non-Frontal:** Face recognition with non-frontal pose is another common problem that has yet to be completely solved and degrades FR performance [41]. The same is true of face recognition with frontal pose, where changes in terms of roll, pitch, and yaw angles impact FR performance as well. Examples of several techniques to handle face recognition across pose are discussed in [41–49]. In [44], pose variations are handled by a method

for reconstructing the virtual frontal view from a given non-frontal face image using Markov random fields and a variant of the belief propagation algorithm. The approach divides the input image into overlapping patches, estimating a globally optimal set of local warps to transform the patches to the frontal view. Oh et al. [47] propose an analytic Gabor feedforward network to handle pose invariance. The network works directly on raw face images using a single sample per identity, and produces directionally projected Gabor magnitude features in the hidden layer. Next, several sets of magnitude features obtained from various orientations and scales are fused in the output layer for classification. The work in [49] handles extreme out-of-plane pose variations. Using their proposed Pose-Aware Models (PAM), face images are processed using several pose-specific deep CNNs. 3D rendering synthesizes multiple face poses from input images to train the models and provide additional robustness to pose variations at test time. Their results show the approach outperforms existing methods evaluated on the IARPA Janus Benchmarks A (IJB-A) and PIPA datasets.

Face image quality factors such as contrast, focus, sharpness, brightness and illumination can also impact FR systems. The work in [50] used a face quality index to show that the filtering of low quality face images can enhance face recognition performance. However, in this work we will only focus on factors 1–3, not image quality. For more on image quality please see [50].

1.2 Our Motivation and Contribution

The goal of our work in this book chapter is to help the forensic operator by improving the process of returning an accurate rank list of potential suspects. We propose a toolkit that can rapidly categorize large databases for several factors that can degrade FR accuracy by detecting facial photos where the subject's eyes are closed, the subject is wearing glasses, or has a non-frontal face pose. The ability to identify these attributes from facial photos in a large database can benefit law enforcement and give operators the option to exclude, group, or enhance these images. An overview of the proposed system is presented in Table 1 where we can see that first, the system input is a mugshot or other face image from the database to be categorized. Then, face detection is performed as well as eye pair detection if possible. Next, HOG features are extracted from the detected face and eyes and each of the three factors are categorized by trained classifiers. The results are then recorded and available for analysis by the examiner.

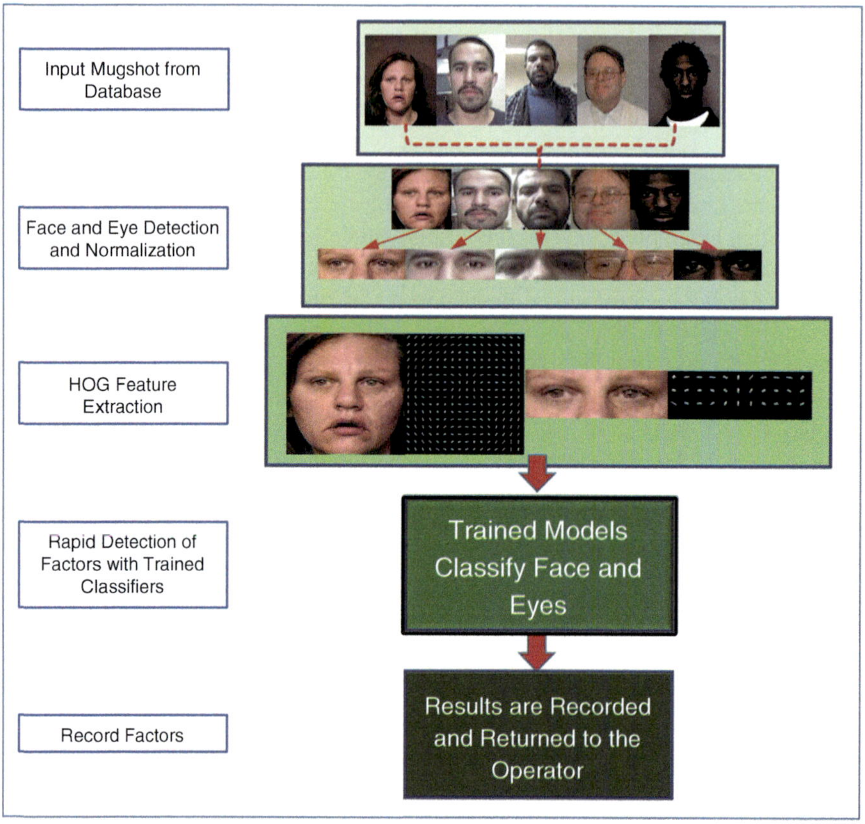

Fig. 1 Overview of our toolkit work flow. *Step 1*: a mugshot image is imported into the interface. *Step 2*: the face and eye pair (if possible) are detected. *Step 3*: HOG features are extracted from the detected face and eyes. *Step 4*: the features are used for classification. *Step 5*: The classification results are recorded and returned to the examiner

2 Methodology

In this section we explain the databases used in our experiments, feature extraction techniques, and the experiments performed to choose the classifiers for our factors.

2.1 Databases

In order to account for the variation in image sizes between each database, we first detect faces using the MTCNN face detector [51] and normalize each face to 130 × 130 pixels. The entire cropped face image is used for classification of the frontal face factor, while the eye pairs from each face are found with a cascade object detector

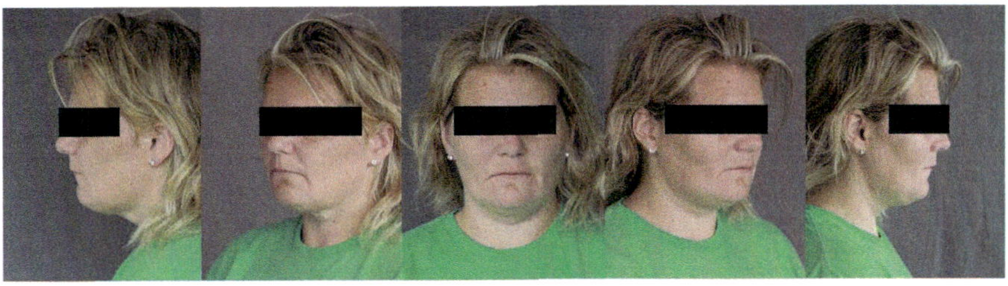

Fig. 2 Sample images from DB1 captured at -90° to $+90^\circ$ poses at 45° intervals. Additionally, every subject has an identical set of images with the eyes closed

Fig. 3 Sample images from DB2 with various poses, backgrounds, and illumination conditions

using the Viola-Jones algorithm [52]. The eye pair is then cropped to 90×30 pixels to classify the eyes and glasses factors.

- **Good Quality Face Database (DB1):** The database contains face images collected indoors at a distance of 2 meters from 1 session with a Canon EOS 5D Mark II and Mark III camera. Images were captured from -90° to $+90^\circ$ poses at 45° intervals, each with the subject's eyes open and closed. Overall, the database is composed of 1719 subjects and 15,240 images. This data closely represents high quality mugshot photos and is therefore used as our baseline database for classification. A sample of these images can be seen in Fig. 2.

- **Multiple Encounter Dataset II (DB2):** This database is a collection of law enforcement submissions of deceased persons with multiple prior encounters. The dataset consists of 518 subjects collected indoors at various profile, near frontal, and frontal poses under variable illumination conditions. The sensors used to capture these images are unknown and result in a wide range of image dimensions, with 70% being approximately 0.3 mega-pixels. The number of samples per subject varies, with 262 of the subjects having 1 sample and the remaining 256 subjects having anywhere between 2 and 18 samples, totaling 1,309 images. This mugshot data represents the range of variations that can be frequently encountered in real world scenarios. A sample of these images can be seen in Fig. 3.

Table 1 Summary of the number of images used in each scenario from every database. A * denotes the addition of augmented data

Databases	Number of images					
	Eyes open	Eyes closed	Glasses	No glasses	Frontal	Non-frontal
Good quality face database	1636	1614	1174	1181	3283	6558
Multiple encounter dataset II	905	904*	287*	336*	940	1077*
Combined database	2541	2518	1461	1517	4223	7635

- **Combined Database (DB3):** This database is the combination of DBs one and two. By combining these databases, we can train classifiers that capture the variance in both high and low quality mugshot submissions.

- **Database Partitioning:** While the experiments are the same across each of the three factors, the data used for each factor is unique. The eyes and glasses classification data in DB1 and DB2 are composed of only frontal face images where both eyes can be detected. In order to compensate for the low number of glasses, closed eyes, and non-frontal face samples in DB2, data augmentation is performed in order to balance the classes. Synthetic data was created to augment the eyes and face factors. For the eyes, all 21 subjects with closed eyes were augmented, creating 42 additional images per subject and 882 images total. Each closed eye pair was flipped along the horizontal axis. Then these two eye pairs, the original and flipped samples, were additionally augmented with Gaussian noise, salt and pepper noise, two levels of increased contrast, two levels of increased and two levels of decreased brightness by changing the gamma parameter, and two increasing levels of Gaussian blur, creating 22 total images. Finally, each of these 22 augmented images is given a random x and y axis translation of $\pm$ 5 pixels. For the face factor, the non-frontal face images were flipped along the x axis and Gaussian blur was added to the original, creating two additional images per sample totaling 718 additional non-frontal face images. Lastly, the glasses factor was supplemented with subjects from the Labeled Faces in the Wild [53] database, containing labeled faces that span a range of in the wild conditions including pose, lighting, race, accessories, occlusion, and background. 280 subjects with glasses and 324 subjects without glasses were used from this database to supplement DB2. A summary of the number of images by database for each factor is shown in Table 1.

2.2 *Feature Extraction*

In this work we tested two common global feature descriptors, Histogram of Oriented Gradients (HOG) [54], and Local Binary Patterns (LBP) [55]. The LBP operator is a texture descriptor that computes patterns in an image by thresholding local neighborhoods, commonly 3 × 3, around every pixel in an image at the central pixel. The resulting possible 256 8-bit patterns are then converted to decimal form. The binary pattern for the pixels in a 3 × 3 neighborhood are computed as follows,

$$LBP(X_c, Y_c) = \sum_{n=o}^{7} h(g_n - g_c)2^n \quad (1)$$

where (X_c,Y_c) is the location of the center pixel *c*, *n* is the number of neighbor pixels, g_n is the grayscale value at pixel *n*, g_c is the grayscale value at *c*, and $h(g_n - g_c)$ is 1 if $h(g_n - g_c) \geq 0$ and 0 otherwise.

HOG features were introduced in [54] for human detection and have been used successfully in a number of applications in object detection and classification. HOG features divide the image into small regions called cells, where a histogram of gradient directions are computed. To make the descriptor more invariant to illumination changes, the histograms are then normalized by accumulating a measure of local histogram energy over larger spatial regions called blocks, the results of which are used to normalize all cells in the block. The combination of all normalized histograms create the final HOG descriptor. A visualization of the HOG descriptor can be seen in Fig. 1 for both a face and eye pair sample.

After several comparisons using both methods we found that HOG features consistently outperformed LBP for every factor, especially on more challenging data. Therefore, in all experiments HOG features were used for classification using a cell size of 8 × 8 pixels and a 2 × 2 block size for the eyes and glasses factors. This created a feature descriptor for each sample of length 720. The cell size for the frontal face descriptor was increased to 16 × 16 and used the same block size in order to reduce each sample dimensionality for training. These descriptors were of length 1764.

2.3 Conventional Models for Classification

We used 23 different models to perform our classification experiments, which including multiple Support Vector Machines [56], K-Nearest Neighbors [57], Decision Trees [58], and Ensemble classifiers [59]. To select the best performing classification models, we perform 10-fold cross-validation on each factor in every database with all available models, creating 9 total scenarios. The results from these experiments allow us to choose the models that will generalize best to classify each of our factors across diverse data.

2.4 Convolutional Neural Networks for Classification

In addition to using the previously mentioned models, we also trained two popular CNNs, AlexNet and GoogLeNet, on DB3 for each of our three classification factors.

AlexNet Architecture AlexNet [60] is an eight layer CNN consisting of five convolutional layers, three fully connected layers, and takes an input image of size 227×227 pixels. The output of the last fully-connected layer is fed to a 1000-way Softmax layer which outputs probabilities for 1000 class labels. For our purposes we use transfer learning and change the last three layers to instead classify 2 labels for each factor, e.g. are the eyes open or closed.

GoogLeNet Architecture GoogLeNet [61] is a 22 layer CNN that takes an input image of size 224×224 pixels and can also classify 1000 class labels. GoogLeNet uses nine Inception modules that convolve 1×1, 3×3, and 5×5 filters in parallel, followed by a 3×3 max pooling. We again change the last three layers of this network to classify 2 labels for each factor.

Training, Testing, and Optimization In the experiments performed with CNNs, DB3 was split using 60% of the data for training, 20% for validation, and 20% for testing each of the three factors. To train the networks we selected a batch size of 100 for the eyes and frontal face factors, and 50 for the glasses factor due to the much smaller amount of available data, and conducted empirical optimization on learning rate, epoch, and momentum parameters, repeating the same process for both networks that resulted in the best classification accuracy for each factor. First, an initial range of eight learning rates (LR) were tested, evenly spaced from 0.01 to 0.0001, holding all other parameters the same. Then a sub-range of learning rates that performed best was selected, and a final set of five evenly spaced LRs were chosen from this subset. Using each of these selected LRs, experiments for every combination of epochs from $4, 8, \ldots, 20$, and momentum of $0.6, 0.65, \ldots, 0.95$ were conducted. An epoch value of 16 worked best for both networks in the eyes and frontal face classifiers, and a value of 12 for both networks for the glasses. AlexNet momentum values of 0.85, 0.9, and 0.85 were the best for eyes, frontal face, and glasses respectively, and 0.95 for every experiment using GoogLeNet.

3 Experimental Results

In this work, we use CNNs, an array of traditional classifiers with different kernel functions, including quadratic, cubic, and Gaussian to perform classification. In our first experiments, we illustrate what models perform the best on our datasets using HOG features with both good quality and challenging data. We also find the models that generalize the best to the combination of those two datasets and can perform well on real world data. In our second experiment, we train and test two CNNs on DB3 to observe any improvements over using HOG features, as well as implement

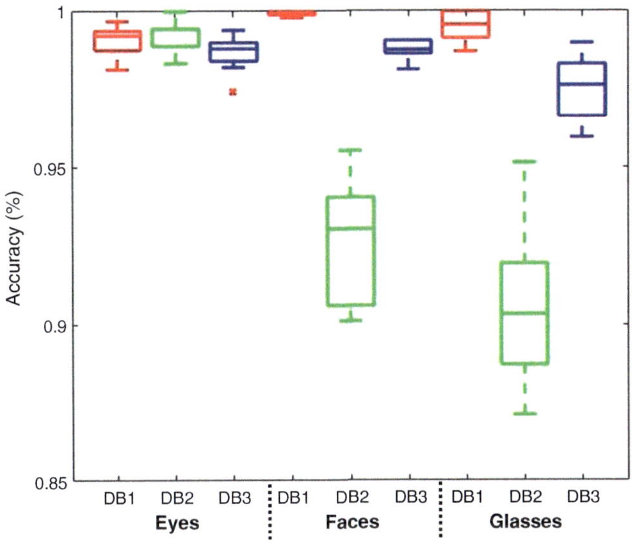

Fig. 4 Classification results for open and closed eyes (Left), frontal and non-frontal faces (Middle), and presence or absence of glasses (Right) on the same axis

score level fusion of the traditional and CNN classifiers by summing the final scores from each class.

3.1 Classification Results

For the classification of eyes, frontal faces, and glasses in Experiment 1 we found that SVMs achieved the best classification results with the exception of a Fine KNN for classifying eyes open or closed in DB2. As expected, accuracy was nearly the same or lower for DB2 in each of the three classification scenarios. This is due to the unconstrained environments in which these images were captured as well as the relatively low number of training images compared to DB1. The classification performance across each database and each factor can be seen in the box plot in Fig. 4. It is important to note that the substantially larger number of training images in DB1 could skew the overall accuracy reported in Table 2 on DB3 by classifying a large number of good quality images and a much smaller number of low quality images. The results in Table 2 show the best achieved accuracy for each of the three DBs, with DB3's columns detailing how accurate that classifier was on DB1 and DB2 data and the final accuracy on DB3. Table 2 also shows the accuracy of the best classifiers trained only on DB1 and only on DB2 data separately. Our results show that the accuracy achieved with the classifier trained on DB3 was nearly identical to the performance when training on each dataset individually. This means that the

Table 2 Summary of the average classification accuracy for each factor using 10-fold cross-validation in all databases. The three DB3 columns show the average cross-validation classification accuracy in terms of the data from DB1 and DB2 individually, as well as the combined databases, DB3, in order to show how well the final trained model can classify both good and challenging data

Average classification accuracy					
			DB3		
Factors	DB1	DB2	DB1	DB2	DB3
Eyes open or closed	99.0	99.2	98.7	98.5	98.7
Frontal or non-frontal faces	99.9	93.5	99.8	93.2	98.8
Glasses present or absent	99.6	89.2	99.6	89.7	97.6

Table 3 Comparison of the average classification accuracy on DB3 from 10-fold cross-validation using SVMs, the best achieved accuracy from parameter optimized CNNs on DB3 test data, and fusion of SVM and CNNs

Accuracy: SVM vs CNN					
Factors	SVM	Alexnet	GoogLeNet	SVM+AlexNet	SVM+GoogLeNet
Eyes open or closed	98.7	99.4	99.2	99.5	99.5
Frontal or non-frontal faces	98.8	98.4	98.7	99.7	99.7
Glasses present or absent	97.6	99.0	99.8	99.9	99.8

best performing classifier of DB3 generalized very well to the combined data and can accurately classify good and poor quality images.

For our second experiment, we optimized AlexNet and GoogLeNet to train and test on DB3 and compare classification accuracy against the SVMs from experiment 1. The results are shown in Table 3. We observed that both CNNs improved classification of open and closed eyes by as much as 0.7% and the glasses factor by over 2%. However, the CNNs for frontal and non-frontal face classification were nearly the same as the SVM. After fusing the scores across all scenarios, we were able to achieve almost 100% accuracy for all 3 factors.

4 Conclusions and Future Work

We investigated the advantages of rapid categorization of factors that impact face recognition performance when processing large scale face datasets collected under constrained and unconstrained conditions. To perform the experiments we used three databases, Good Quality Face Dataset, Multiple Encounter Dataset II, and a combination of the two. We propose a software toolkit that uses multiple trained classifiers to classify face images as frontal or non-frontal, eyes open or closed, and presence or absence of glasses. To perform this classification we trained a variety of algorithms with 10-fold cross-validation, including SVMs using LBP and HOG features as well as two well known convolutional neural networks.

Several scenarios were trained for each factor, testing 23 different conventional models with a number of kernel functions in order to select the model and kernel function combination that best classified each factor. CNNs were optimized to find the ideal parameters for training and testing. Our experimental results show that our models were able to classify each factor in our most challenging database at least 90% of the time, and over 99% for all factors in DB3 after implementing score level fusion. The most challenging factor was frontal and non-frontal faces from DB2. This is likely due to the subjective nature of the labeling process of these face images as either frontal or non-frontal because many of them were very close to being in either class. The same can be said of the eyes open or closed data, where a majority of the misclassified samples were eyes that were very slightly open. When combined with variations in expression, lighting, distance, and background, many of these samples proved to be quite challenging to classify correctly.

Based on our results we conclude that a toolkit which almost simultaneously classifies several well-known factors that affect facial recognition systems can be very beneficial to law enforcement and forensic operators at identifying individuals in the gallery. The use of hand crafted features with well-known models such as SVMs and popular CNNs can quickly find and categorize well over 90% of face images in a large database correctly, raising the overall quality of images to match against the gallery by excluding or grouping poor quality faces, or even enhancing them. In the future, we intend to improve this work by including additional factors that affect FR performance.

References

1. Jain AK, Ross A (2015) Bridging the gap: from biometrics to forensics. Philos Trans R Soc B 370(1674):20140254
2. Tistarelli M, Grosso E, Meuwly D (2014) Biometrics in forensic science: challenges, lessons and new technologies. In: International workshop on biometric authentication. Springer, pp 153–164
3. Meuwly D, Veldhuis R (2012) Forensic biometrics: from two communities to one discipline. In: 2012 BIOSIG-proceedings of the international conference of the biometrics special interest group (BIOSIG). IEEE, pp 1–12
4. Dessimoz D, Champod C (2008) Linkages between biometrics and forensic science. In: Handbook of biometrics. Springer, Boston, pp 425–459
5. Champod C, Tistarelli M (2017) Biometric technologies for forensic science and policing: state of the art. In: Handbook of biometrics for forensic science. Springer, pp 1–15
6. Jain AK, Klare B, Park U (2011) Face recognition: some challenges in forensics. In: 2011 IEEE international conference on automatic face & gesture recognition and workshops (FG 2011). IEEE, pp 726–733
7. Jain AK, Klare B, Park U (2012) Face matching and retrieval in forensics applications. IEEE Multimed 19(1):20
8. Lee J, Jain A, Tong W et al (2012) Image retrieval in forensics: tattoo image database application. IEEE MultiMed 19(1):40–49
9. Wing BJ (2013) The ANSI/NIST-ITL standard update for 2011 (data format for the interchange of fingerprint, facial and other biometric information). Int J Biom 5(1):20–29

10. Bourlai T (2016) Face recognition across the imaging spectrum. Springer, Switzerland
11. Bourlai T, Narang N, Cukic B, Hornak L (2012) On designing a swir multi-wavelength facial-based acquisition system. In: Infrared technology and applications XXXVIII, vol 8353. International Society for Optics and Photonics, p 83530R
12. White D, Dunn JD, Schmid AC, Kemp RI (2015) Error rates in users of automatic face recognition software. PLoS One 10(10):e0139827
13. Jain AK, Li SZ (2011) Handbook of face recognition. Springer, New York
14. Ding X, Wang L (2011) Facial landmark localization. In: Handbook of face recognition. Springer, New York, pp 305–322
15. Whitelam C, Bourlai T (2015) Accurate eye localization in the short waved infrared spectrum through summation range filters. Comput Vis Image Underst 139:59–72
16. El-Sayed MA, Khafagy MA (2014) An identification system using eye detection based on wavelets and neural networks. 1401.5108
17. Bourlai T, Jafri Z (2011) Eye detection in the middle-wave infrared spectrum: towards recognition in the dark. In: 2011 IEEE international workshop on information forensics and security. IEEE, pp 1–6
18. Dutta A, Günther M, El Shafey L, Marcel S, Veldhuis R, Spreeuwers L (2015) Impact of eye detection error on face recognition performance. IET Biom 4(3):137–150
19. Du C, Su G (2005) Eyeglasses removal from facial images. Pattern Recogn Lett 26(14):2215–2220
20. Taylor M (2017) The art of facial recognition. https://www.forensicmag.com/article/2017/03/art-facial-recognition. Accessed 18 Feb 2018
21. Mandal B, Li L, Wang GS, Lin J (2017) Towards detection of bus driver fatigue based on robust visual analysis of eye state. IEEE Trans Intell Transp Syst 18(3):545–557
22. González-Ortega D, Díaz-Pernas FJ, Antón-Rodríguez M, Martínez-Zarzuela M, Díez-Higuera JF (2013) Real-time vision-based eye state detection for driver alertness monitoring. Pattern Anal Appl 16(3):285–306
23. Zhao L, Wang Z, Zhang G, Qi Y, Wang X (2017) Eye state recognition based on deep integrated neural network and transfer learning. Multimed Tools Appl 77(15):1–24
24. Kim KW, Hong HG, Nam GP, Park KR (2017) A study of deep CNN-based classification of open and closed eyes using a visible light camera sensor. Sensors 17(7):1534
25. Ji Y, Wang S, Lu Y, Wei J, Zhao Y (2018) Eye and mouth state detection algorithm based on contour feature extraction. J Electron Imaging 27(5):051205
26. Hassan H, Yaacob S, Radman A, Suandi SA (2016) Eye state detection for driver inattention based on Lucas Kanade optical flow algorithm. In: 2016 6th international conference on intelligent and advanced systems (ICIAS). IEEE, pp 1–6
27. Radlak K, Smolka B (2012) A novel approach to the eye movement analysis using a high speed camera. In: 2012 2nd international conference on advances in computational tools for engineering applications (ACTEA). IEEE, pp 145–150
28. Fogelton A, Benesova W (2016) Eye blink detection based on motion vectors analysis. Comput Vis Image Underst 148:23–33
29. Pauly L, Sankar D (2015) Detection of drowsiness based on hog features and SVM classifiers. In: 2015 IEEE international conference on research in computational intelligence and communication networks (ICRCICN). IEEE, pp 181–186
30. Dong Y, Zhang Y, Yue J, Hu Z (2016) Comparison of random forest, random ferns and support vector machine for eye state classification. Multimed Tools Appl 75(19):11763–11783
31. Eddine BD, Dos Santos FN, Boulebtateche B, Bensaoula S (2018) Eyelsd a robust approach for eye localization and state detection. J Signal Process Syst 90(1):99–125
32. Sharif M, Bhagavatula S, Bauer L, Reiter MK (2016) Accessorize to a crime: real and stealthy attacks on state-of-the-art face recognition. In: Proceedings of the 2016 ACM SIGSAC conference on computer and communications security. ACM, pp 1528–1540
33. Ying H, Da Su G, Chen JS (2014) Automatic eyeglasses removal of frontal facial images for face recognition. Appl Mech Mater 571:697

34. Alorf A, Abbott AL (2017) In defense of low-level structural features and SVMs for facial attribute classification: application to detection of eye state, mouth state, and eyeglasses in the wild. In: 2017 IEEE international joint conference on biometrics (IJCB). IEEE, pp 599–607
35. Mohammad AS, Rattani A, Derahkshani R (2017) Eyeglasses detection based on learning and non-learning based classification schemes. In: 2017 IEEE international symposium on technologies for homeland security (HST). IEEE, pp 1–5
36. Zhao M, Zhang X, Shi Z, Chen T, Zhang F (2018) Eyeglasses detection, location and frame discriminant based on edge information projection. Multimed Tools Appl 77(12):14931–14949
37. Lazarus MZ, Gupta S (2016) A low rank model based improved eye detection under spectacles. In: Ubiquitous computing, electronics & mobile communication conference (UEMCON), IEEE annual. IEEE, pp 1–6
38. Basbrain AM, Al-Taie I, Azeez N, Gan JQ, Clark A (2017) Shallow convolutional neural network for eyeglasses detection in facial images. In: Computer science and electronic engineering (CEEC), 2017. IEEE, pp 157–161
39. Wang Y, Ou X, Tu L, Liu L (2018) Effective facial obstructions removal with enhanced cycle-consistent generative adversarial networks. In: International conference on artificial intelligence: methodology, systems, and applications. Springer, pp 210–220
40. Zhang X, Pham DS, Venkatesh S, Liu W, Phung D (2015) Mixed-norm sparse representation for multi view face recognition. Pattern Recognition 48(9):2935–2946.
41. Ding C, Xu C, Tao D (2015) Multi-task pose-invariant face recognition. IEEE Trans Image Process 24(3):980–993
42. Zhang X, Gao Y (2009) Face recognition across pose: A review. Pattern Recogn 42(11):2876–2896
43. Zhang X, Pham D-S, Venkatesh S, Liu W, Phung D (2015) Mixed-norm sparse representation for multi view face recognition. Pattern Recogn 48(9):2935–2946
44. Ho HT, Chellappa R (2013) Pose-invariant face recognition using Markov random fields. IEEE Trans Image Process 22(4):1573–1584
45. Xu W, Shen Y, Bergmann N, Hu W (2018) Sensor-assisted multi-view face recognition system on smart glass. IEEE Trans Mobile Comput 17(1):197–210
46. Shao M, Zhang Y, Fu Y (2018) Collaborative random faces-guided encoders for pose-invariant face representation learning. IEEE Trans Neural Netw Learn Syst 29(4):1019–1032
47. Oh B-S, Toh K-A, Teoh ABJ, Lin Z (2018) An analytic gabor feedforward network for single-sample and pose-invariant face recognition. IEEE Trans Image Process 27(6):2791–2805
48. Deng W, Hu J, Wu Z, Guo J (2017) Lighting-aware face frontalization for unconstrained face recognition. Pattern Recogn 68:260–271
49. Masi I, Chang F-J, Choi J, Harel S, Kim J, Kim K, Leksut J, Rawls S, Wu Y, Hassner T et al (2018) Learning pose-aware models for pose-invariant face recognition in the wild. IEEE Trans Pattern Anal Mach Intell 41(2):379–393
50. Abaza A, Harrison MA, Bourlai T, Ross A (2014) Design and evaluation of photometric image quality measures for effective face recognition. IET Biom 3(4):314–324
51. Zhang K, Zhang Z, Li Z, Qiao Y (2016) Joint face detection and alignment using multitask cascaded convolutional networks. IEEE Signal Process Lett 23(10):1499–1503
52. Viola P, Jones M (2001, December) Rapid object detection using a boosted cascade of simple features. In: Proceedings of the 2001 IEEE computer society conference on computer vision and pattern recognition, 2001. CVPR 2001, vol 1. IEEE, pp I–I
53. Huang GB, Mattar M, Berg T, Learned-Miller E (2008) Labeled faces in the wild: a database for studying face recognition in unconstrained environments. In: Workshop on faces in'Real-Life'Images: detection, alignment, and recognition
54. Dalal N, Triggs B (2005) Histograms of oriented gradients for human detection. In: IEEE computer society conference on computer vision and pattern recognition, 2005. CVPR 2005, vol 1. IEEE, pp 886–893
55. Ojala T, Pietikäinen M, Harwood D (1996) A comparative study of texture measures with classification based on featured distributions. Pattern Recogn 29(1):51–59
56. Cortes C, Vapnik V (1995) Support-vector networks. Mach Learn 20(3):273–297

57. Fix E, Hodges JL Jr (1951) Discriminatory analysis-nonparametric discrimination: consistency properties. Technical report, California Univ Berkeley
58. Breiman L (2017) Classification and regression trees. Routledge, Belmont
59. Dietterich TG (2000) Ensemble methods in machine learning. In: International workshop on multiple classifier systems. Springer, pp 1–15
60. Krizhevsky A, Sutskever I, Hinton GE (2012) Imagenet classification with deep convolutional neural networks. In: Advances in neural information processing systems, pp 1097–1105
61. Szegedy C, Liu W, Jia Y, Sermanet P, Reed S, Anguelov D, Erhan D, Vanhoucke V, Rabinovich A (2015) Going deeper with convolutions. In: Proceedings of the IEEE conference on computer vision and pattern recognition, pp 1–9

Classification of Soft Biometric Traits When Matching Near-Infrared Long-Range Face Images Against Their Visible Counterparts

Neeru Narang and Thirimachos Bourlai

Abstract In this chapter, we investigate the advantages and limitations of the heterogeneous problem of matching Near-Infrared (NIR) long-range, night time, face images against their visible counterparts. Image quality degradation can result due to a variety of factors including low illumination, variable standoff distance, and is responsible for performance degradation of conventional face recognition (FR) systems. In addition to intra-spectral matching (i.e. NIR vs. NIR face images), cross-spectral matching (i.e. matching NIR face images against their visible counterparts) is a challenging matching scenario that increases system complexity. In this work, we propose the usage of a set of FR algorithms when working with operational-based face matching scenarios, namely, where the face images used are collected by a night vision, long range (from 30 to 120 m), NIR-based face imaging system. First, we establish a system identification baseline using a set of commercial and academic face matchers. To improve baseline performance, we propose a scenario dependent convolutional neural network (CNN) to, first, categorize the face images of our challenging face dataset, in terms of gender, ethnicity, and facial hair. For each of the aforementioned generated categories, we apply our proposed algorithmic pipeline including, image restoration and a multi-feature based fusion scheme. Then, a set of FR algorithms are used before and after image restoration and data categorization. Based on the experimental results, we conclude that our proposed image restoration and fusion schemes, as well as the usage of demographic-based face categories, result in improved identification performance. For example, for the 30 m vs. 30 m NIR face matching scenario, the rank-1 identification rate is improved from 48% (all vs. all) using a commercial face matching system to 73% (all vs. all) and to 82% (if

N. Narang
Lane Department of Computer Science and Electrical Engineering, West Virginia University, Morgantown, WV, USA
e-mail: nneeru@mix.wvu.edu

T. Bourlai (✉)
Lane Department of Computer Science and Electrical Engineering, Multispectral Imagery Lab (MILab), West Virginia University, Morgantown, WV, USA
e-mail: Thirimachos.Bourlai@mail.wvu.edu; ThBourlai@mail.wvu.edu

T. Bourlai et al. (eds.), *Securing Social Identity in Mobile Platforms*, Advanced Sciences and Technologies for Security Applications,
https://doi.org/10.1007/978-3-030-39489-9_5

we use only the male with beards face data category). Experimental results suggest that our proposed methodological approach can improve system performance (i.e. efficiently identifying the subject of interest) on various cross-spectral face matching scenarios.

1 Introduction

The existing standard face recognition (FR) systems result in high identification rates when operating in controlled conditions, e.g. indoors, during day time and at short standoff distances [1, 2]. However, in law enforcement and security applications, investigators deal with mixed FR scenarios that involve matching probe face images captured by different camera sensors, under un-controlled conditions (e.g. the subject is far away from the camera, outdoors and at night time conditions) against the good quality face images (e.g. mug shots), acquired using high definition camera sensors (e.g. DSLR cameras) [3]. Under such challenging conditions, conventional face recognition algorithms (using handcraft features such as Local Binary Patterns, HOGs etc.) often provide unsatisfactory results. Thus, efficient and reliable FR systems need to be developed, which are capable of supporting law enforcement officers establish high identification rates under all conditions. While modern approaches (e.g. based on convolutional neural networks) are promising, one complementary solution is to use demographic information, namely age, gender, and ethnicity so that face images are grouped before matching is performed [4]. These traits are also referred to as soft biometric traits, which provide several advantages to operational capabilities of traditional biometric systems [5].

Soft biometric traits, have been regularly used by the biometrics research community in various applications [6–8]. For example, Jain et al. [5], utilized soft traits such as ethnicity, weight, gender and height before matching and concluded that they can significantly improve the recognition performance of fingerprint or other biometric systems. When performing the automatic classification of soft traits in controlled conditions, e.g. indoors, outdoors, during day time, at short ranges etc., the capabilities of the existing classification systems can result in classification rates of acceptable performance. However, automatic classification when working under un-controlled conditions e.g. face images collected outdoors, at night time conditions, and when the subject is far away from the camera, is a very interesting and challenging research topic. What follows is a discussion on the goals and contributions of this chapter.

1.1 Goals and Contributions

In our original work found in [2] we report the rank-1 identification accuracy results of an inter-distance and intra spectral FR system, which was designed to collect face images at large standoff distances (30, 60, 90 and 120 m) in both daytime and nighttime conditions. In [9], we extended that work and focused on a cross-distance and cross-spectral face recognition system. There we reported the rank-1 identification rates before and after demographic grouping of the data (grouping was performed manually) into a male and a female class.

In this work we continue investigating some of the challenges of un-constrained face recognition by focusing on a set of FR scenarios. We use a dual-band (visible and NIR) private face database collected at night time conditions and variable standoff distances, ranging from 30 to 120 m, in a 30 m interval as shown in Fig. 1. We perform both intra-spectral and cross-spectral face matching experiments, before and after using demographic grouping of this face database. Thus, we propose a deep learning based, scenario-dependent, and band-adaptable (it is working well for both visible and NIR face images) algorithmic approach for the automated classification of a set of soft biometric traits. These include the following classes, namely, male vs. female, followed by a male class with and without having a beard. The female class is also sub-categorized into an Asian and a Caucasian class. Since we are dealing with low quality face images, in our approach, we take advantage of an image restoration approach, which improves the quality of distorted, long range NIR face images. Then, we use a multi-feature based fusion scheme: first, Gabor Wavelets, Histogram of Gradients (HOG) and Local Binary Pattern (LBP) based feature descriptors are empirically selected and, then, a set of fusion score level schemes are proposed to improve FR performance.

In this work, both commercial and academic face matchers are used and a set of experiments is performed, indicating that our proposed image restoration, fusion schemes and the usage of demographic information of the database, achieves significantly better performance results than the established baseline, e.g. for the 30 m vs. 30 m NIR face matching scenario, the rank-1 identification rate is improved from 48% (all vs. all) using a commercial face matching system to 73% (all vs. all) and to 82% (if we use only the male with beards face data category).

To our knowledge, this is the first time a VIS-NIR face recognition system is evaluated using such an approach, using demographic grouping and a set of image restoration, fusion and face matching algorithms. Also, there is no reported work that evaluates the impact of demographic grouping (in terms of ethnicity, gender and facial hair) in a cross-spectral (VIS against NIR) FR system performance.

2 Background

Liu et al. [10], proposed a method to synthesize visible from NIR face images. The authors reported that there was significant improvement in the face verification results after image synthesis. Chen et al. [11], proposed a method of synthesizing the VIS from NIR images based on patch based transformation method. Klare et al. [12], performed cross-spectral matching between NIR and VIS face images for the database collected at a short distance of 0.7 m under controlled environmental conditions as represented in Table 1. In [2], a NIR sensor was used to capture face images at long-range stand-off distances and at night time conditions. All NIR face

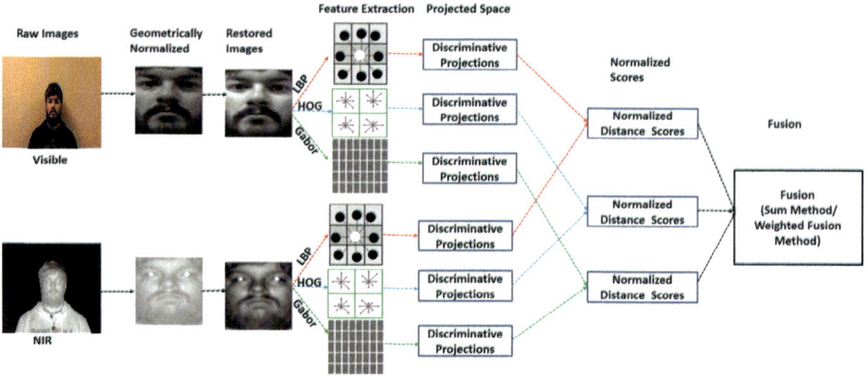

Fig. 1 Overview of our proposed VIS – NIR face recognition system

images were acquired at four different standoff distances (30, 60, 90 and 120 m). Using the face datasets generated for each standoff distance, intra distance and cross-spectral face matching experiments were performed.

Maeng et al. [13], performed cross distance and cross-spectral matching between VIS and NIR band (VIS 1 m and NIR 60 m). In [14], cross distance and cross-spectral matching for short and long range distances was performed, in both daytime and night time environments (VIS 1 m and NIR at 60, 100 and 150 m). In [14], the authors used the SIFT descriptors for feature extraction and the LDA based subspace to minimize the intra-subject differences due to the modality difference. They used the CASIA HFB (NIR-VIS) database at a distance of 1.2 m to train the system and long distance heterogeneous face (LDHF)-DB long range distance data for testing. Yi et al. [15], proposed a learning based approach, where a canonical correlation analysis (CCA) based method was used in order to learn the correlation between NIR and VIS images (image synthesis). In that study, the authors used the short distance (1.5 m) indoor, visible band, face database as the gallery set. Similar work on image synthesis but when operating in the MWIR vs. the NIR band was reported in [16, 17].

Kang et al. [24], proposed an image restoration method for cross-spectral matching. The developed algorithm was trained using low-quality face images (150 m NIR) and their corresponding high-quality face images (1 m NIR). Omri et al. [22], proposed a method to fuse the images collected under different spectral bands (VIS and NIR) at the score level. They further extended their work to a long range face database collected under challenging conditions at 60, 100 and 150 m distances [23].

Most of the work reported in the literature in the area of heterogeneous face matching in the VIS and NIR bands, is based on learning subspace methods. To generate the subspace [1, 2, 10, 12, 13, 24], good quality face images, collected at a short standoff distance (i.e. 1.0 m in both VIS and NIR bands) were used for training and long distance data (60, 100 and 150 m) was used for testing as

Table 1 Previous work related to heterogeneous face matching

Related papers	Distance range	Cross spectral matching approach	Accuracy results %
	Short range	**Synthesis based**	
Liu et al. [10]	VIS – NIR (0.5 m ∼ 1.2 m)	CCA	30.0 (ROC)
Wang et al. [18]	VIS – NIR (∼ 1.5 m)	Face Analogy	95.0 (ROC)
Zhang et al. [19]	VIS – NIR (∼ 1.5 m)	Sparse Representation	99.5 (CMC)
Chen et al. [11]	VIS – NIR (∼1.5 m)	Learning Mapping	97.3 (CMC)
		Subspace based	
Klare et al. [12]	VIS – NIR (∼ 0.7 m)	HOG-LBP	97.06 (ROC)
Yi et al. [15]	VIS – NIR (∼1.5 m)	PCA/LDA/CCA	85.5 (ROC)
Goswami et al. [20]	VIS – NIR (∼1.5 m)	LBPH and LDA	89.5 (CMC)
Raghavendra et al. [21]	VIS – NIR (∼1.5 m)	Particle Swarm Optimization	98.25 (ROC)
Omri et al. [22]	VIS – NIR (∼1.5 m) ∼1.5 m	DWT multispectral	96.00 (ROC)
	Long range	**Fusion and subspace based**	
Maeng et al. [13]	VIS – NIR (1 m, 60 m)	DoG-SIFT	28.0 (CMC)
Omri et al. [23]	VIS – NIR (1 m vs. 60 m)	DWT	44.0 (ROC)
	VIS – NIR (1 m vs. 100 m),	DWT	20.0 (ROC)
	VIS – NIR (1 m vs. 150 m),	DWT	15.0 (ROC)
Maeng et al. [14]	VIS – NIR (1 m vs. 60 m)	SIFT-LBP	81.0 (ROC)
	VIS – NIR (1 m vs. 100 m),	SIFT-LBP	61.0 (ROC)
	VIS – NIR (1 m vs. 150 m),	SIFT-LBP	20.0.0 (ROC)
Kang et al. [24]	VIS – NIR (1 m vs. 60 m)	Dictionary: LLE	80.0 (ROC)
	VIS – NIR (1 m vs. 100 m),	Dictionary: LLE	70.0 (ROC)
	VIS – NIR (1 m vs. 150 m),	Dictionary: LLE	33.0.0 (ROC)
Bourlai et al. [2]	VIS – NIR (30 m vs. 30 m),	CSU	98.0 (CMC)
	VIS – NIR (60 m vs. 60 m),	CSU	91.0 (CMC)
	VIS – NIR (90 m vs. 90 m),	CSU	90.0 (CMC)
	VIS – NIR (120 m vs. 120 m)	CSU	87.0 (CMC)

represented in Table 1. However, in the case of operational FR scenarios, we are not always provided with good quality probe and query face images. Thus, in this work we propose a new image restoration method to solve the problem of dealing with low quality face images captured at long ranges as shown in Fig. 1. The method is based on image de-noising and super-resolution techniques (see Sect. 3) and when combined with demographic grouping and face matching algorithms, the baseline rank-1 identification rate is improved.

3 Methodology

In this section, we outline the system design set-up developed for the collection of our face database. We first discuss, the algorithm proposed for the automated extraction of demographic information. Then, we provide details on the, restoration of low quality face images and the selection of fusion schemes (sum and weighted fusion). Finally, vis analysis we decide which of the approaches has the most significant impact in cross-spectral recognition performance.

3.1 System Design Set Up and Database Collected

A visible and a NIR camera imaging system was used in our live subject capture setup: Canon EOS 5D Mark II is used to collect standard RGB, ultra-high-resolution frontal pose face images in the VIS spectrum. The mid-range NIR camera imaging system is used, provided by Vumii Imaging Inc., which operates at 850 nm. Four standoff distances (30, 60, 90 and 120) were considered to collect face images (see Fig. 2). The database was collected outdoors, at night time, spanning over a time period of 20 days. Recordings of the faces of the subjects were taken with the mid-range camera. Then, the subjects' mug shots were taken using the VIS camera in an indoor controlled environment. In total, 103 subjects (69 male + 34 female) participated in this experiment, and the database included video sequences of full frontal mid-range NIR and VIS faces of different subjects, resulting in a total of 103×5 videos (103×4 NIR outdoors and 103 VIS indoors) per subject.

3.2 Proposed CNN Network for Automatic Prediction of Demographic Information

The classification is performed for two scenarios including, (i) Intra-spectral, where the visible images are selected for both the training and testing, (ii) Cross-spectral, where the visible images are selected for training and NIR images for testing. To perform the intra-spectral and cross-spectral classification, we selected the visual geometry group (VGG) CNN architecture [26]. The deep learning model consists of a number of convolutional layers, max pooling layers and rectified linear unit (ReLU) activation layer along with fully connected layers [26–28]. The output of fully connected layer is fed to a soft-max layer that assign a label to each class, i.e. it will assign a label male or female in terms of gender and Asian or Caucasian in terms of ethnicity class.

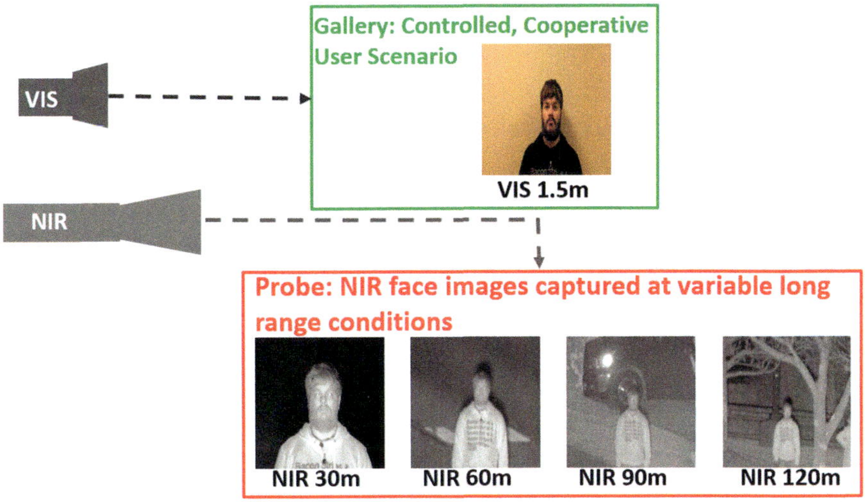

Fig. 2 The live subject-capture setup using the VIS and mid-range NIR cameras. At the bottom we can see a set of face image samples acquired by our system at a night time environment and at variable standoff distances. (Reproduced from Narang et al. [25])

3.2.1 Training and Testing

In the experiments performed, the subjects in the training and test sets are different, and the images were captured at different locations, days and times. We based our work on our previous original approach [29], where we proposed CNN based gender and ethnicity classification approach. In this chapter, we extended our original work to further classify the gender class with the most significant attributes within that class. For the male class, we selected the facial hair as the most discriminating attribute and sub-grouped our male class into a male with or without beard classes. The female class is also further sub-grouped into a female Asian or a female Caucasian class.

Training Data For the limited amount of training data, CNN pre-trained on large databases (ImageNets) is used by the researchers for the recognition and classification tasks. To collect a large training database from available image repositories and to label the database manually is time consuming process. There was no pre-trained multi-sensor network model available to use for our CNN network therefore we trained the models for our original database. To train the system, we have selected 9 different types of databases collected indoors, outdoors, with different camera sensors, expressions, ethnicity, locations and times as shown in Fig. 3.

Testing Data To evaluate the performance of our system, we selected the database collected in our lab, which includes a short distance (1.5 m) visible band face

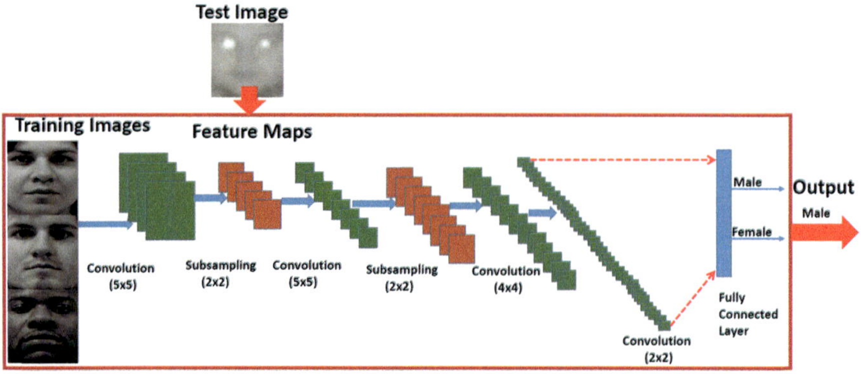

Fig. 3 An overview of our proposed CNN architecture applied to our male vs. female classification problem

dataset and a NIR face dataset, collected at night time and at various stand-off distances (30 to 120 m) (as shown in Fig. 3). Two scenarios are selected to perform the classification including:

Scenario 1: All 9 face databases available are selected to train our system, including, the WVU visible-thermal profile face (VTPF) database, WVUM database [30], Tinders Database [31], QFire FEI [32] and the Libor Spacek Facial database [33]. Our system is finally tested using our own WVU database discussed above.

Scenario 2: A 25% of our (VIS 1.5 m) database and all the 9 databases are selected to train the system, where the rest of our dual band, multi-distance database for the testing without any subjects overlap.

3.3 *Face Identification Steps*

After the demographic grouping of the data using the proposed CNN method, a face database is selected to perform the face matching experiments including, image pre-processing, feature extraction and matching, as discussed in the following subsections.

3.3.1 Pre-processing of Images

To remove the difference in appearance between face images captured in NIR and VIS band and to improve the quality of images, an image restoration approach is

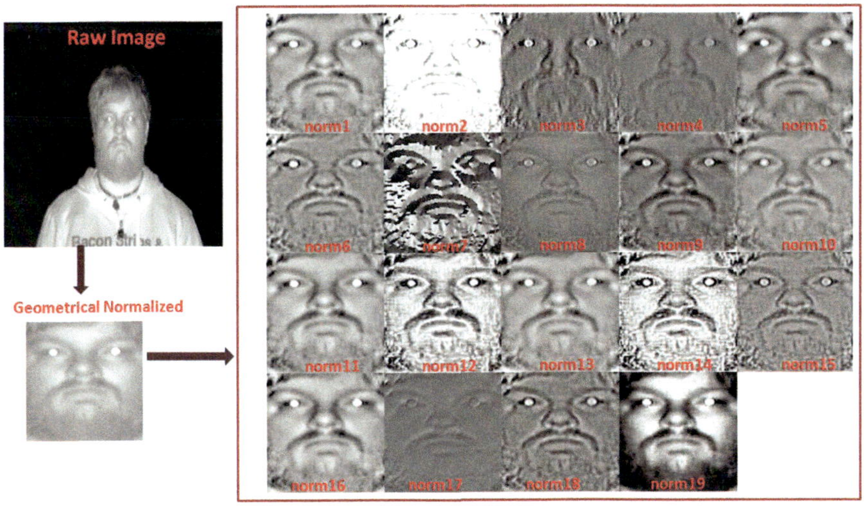

Fig. 4 Raw image (top left), geometrically normalized image (bottom left) and photo-metrically normalized images from a selected set of normalization techniques [36] (right)

proposed. The pre-processing of images is composed of three main steps including: geometrical normalization (*GN*), image restoration and photometric normalization (*PN*). Geometrical Normalization compensates for slight perturbations in the frontal pose. It included two main steps, eye detection and affine transformation and all the faces are canonicalized to the same dimension of 128×128 pixels. In image restoration, the good quality face images are reconstructed from an image enhancement method using a de-noising (*DN*) and a super-resolution (*SR*) method. Super-Resolution is performed based on an example-based *SR* and a *DN* method discussed in [34] and [35] respectively. Finally, *PN* is used to compensate for illumination variations. The advantage of the proposed pre-processing steps is that they can eliminate the irrelevant information while, still preserving face appearance details that are required for face recognition.
In this work, we empirically investigated 19 different *PN* techniques [36] including: adaptive non local means (norm 1), adaptive single scale retinex (norm 2) etc. (as shown in Fig. 4). To optimize the set up (for each scenario), the *PN* method that provides us with the best performance results is selected. The performance results with rank-1 identification rate are shown in our experimental results section (see Figs. 6 and 7).

3.3.2 Combination of Pre-processing Methods

The main challenge is to select the best order of combinations for pre-processing, as we cannot randomly select either *PN*, *DN* or *SR*. To deal with this problem, we selected five combinations such as *comb1 (GN+PN)*, *comb2 (GN+DN+PN)*,

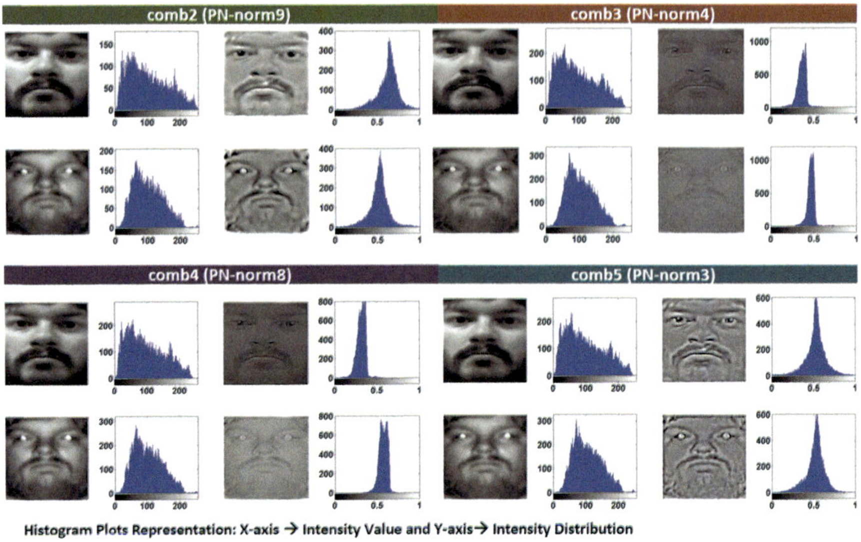

Fig. 5 *PN* techniques applied to restored images (using our proposed image restoration method)

comb3 (GN+SR+PN), *comb4 (GN+DN+SR+PN)* and *comb5 (GN+SR+DN+PN)*. The selection the order of each of pre-processing techniques is critical and also of key importance in improving cross-spectral matching performances in terms of rank-1 identification rate. The results for *LBP, LTP, Gabor, HOG* is shown in our experimental results section in Figs. 6 and 7.

First, we performed a set of experiments including: applying *PN* directly on raw images *comb1* and then on restored images *(DN and SR)*. In Fig. 5, we see histogram plots for gallery (VIS) and probe image (NIR), where *PN* is implemented to the restored face images. the intensity values for gallery and probe images. Whereas, this variation is minimized when we applied the *PN* on the restored images as shown in Fig. 5 and specifically for restored images using *comb3* and *comb4*.

3.3.3 Face Matching

This section will describe the method used for the matching of NIR images to their VIS counterparts. The methods used to represent the face images, to reduce the dimensionality of feature space, and adopted decision level scheme to fuse the features scores to perform the matching. In face recognition, discriminant based approaches provide us with high performance results specifically for the database collected under different illumination conditions [37]. Our work is unique when compare to the aforementioned paper in three different ways: (i) In terms of our *image restoration* approach, (ii) In terms of our feature extraction (FE) approach composed of three different FE methods, namely, Gabor, LBP and HOG. (iii) In

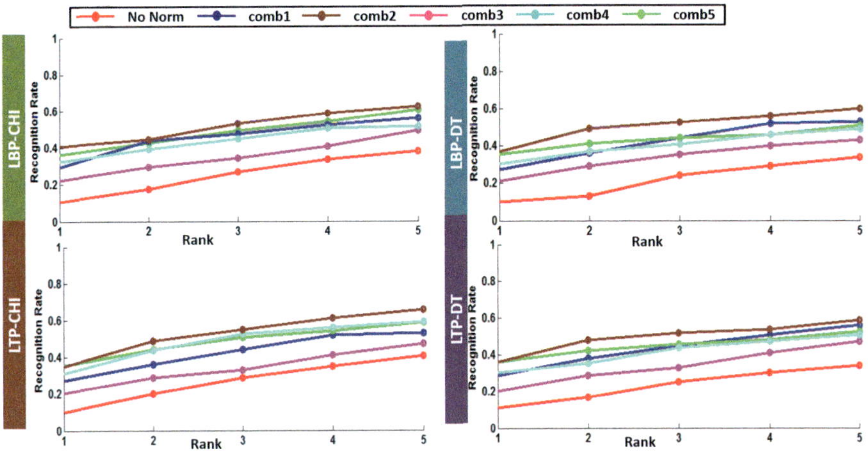

Fig. 6 CMC curves comparing the identification performance (first 5 ranks) for cross-spectral matching (Gallery-VIS and Probe-NIR): LBP-CHI (top-left), LBP-DT (top-right), LTP-CHI (bottom-left) and LTP-DT (bottom-right). Five pre-processing combinations selected such as *comb1*, *comb2*, *comb3*, *comb4* and *comb5*

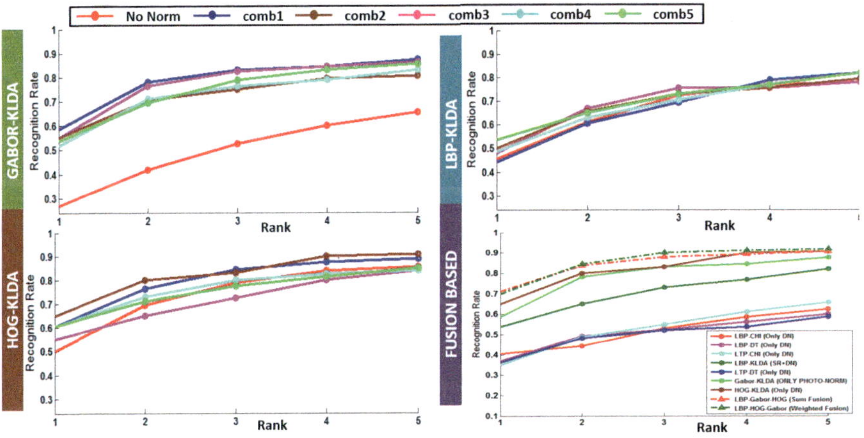

Fig. 7 CMC curves comparing identification performance (for rank to rank 5) results, for cross-spectral matching, for five selected combinations of pre-processing techniques: LBP/KLDA (top-right), GABOR/KLDA (top-left) and HOG/KLDA (bottom-left)

terms of our proposed fusion scheme, where we tested about 11 different fusion-based scenarios, e.g. sum fusion, weighted fusion etc. We also developed and tested the FU-GLBHG face matching method, based on the fusion (FU) of Gabor, LBP and HOG features.

– **Feature Extraction** Facial representation is defined as the description of facial information, to transform the matrix array for face into column vector. Global and local descriptors are extracted. We selected three feature descriptors Local Binary Patterns (LBP), Gabor wavelets and histogram of gradients (HOG). LBP is a gray scale-invariant local image descriptor and used to get the appearance and texture information [9]. Gabor method is used to detect high frequency components (edges) in different images. HOG has proved as one of the successful local shape descriptor and significantly outperforms the available feature sets including wavelets [38]. The main idea is to estimate the local histograms of image gradient orientation in a dense grid.

- **Feature Subspace** After the extraction of features based on LBP, Gabor and HOG methods in both the cross-spectral bands (NIR and VIS images), the main challenge was to reduce the dimensionality of the feature space. We adopted the method used by Tan et al. [37]. They used LBP and Gabor image feature descriptors. We extended our work to LBP, Gabor and HOG and used KLDA learning based analysis for the good quality face images restored from our proposed image restoration method.

- **Matching** For the training phase, we selected both VIS (1 m) and long range NIR images (30 m). Matching is performed only for non-overlapping subjects, those not present in the training set. To perform matching on test set, we apply the following steps including pre-processing, to restore the good quality face images, and feature extraction.

For a given input face image (gallery/probe), its Gabor, HOG and LBP features are extracted and separately projected to the optimal discriminant feature space. The features extracted from the visible band face images (gallery set) [37] and probe face images are projected to feature space. The distance scores are measured using the cosine distance score method.

To normalize the scores, well known methods are available namely the following, z-score, min, max, tanh etc. [22, 39]. In this work, score normalization is performed using the $z\text{-}score$ (score values lie between 0 and 1). The normalized scores z_{LBP}, z_{HOG} and z_{GABOR} are fused based on a decision score level method. In most research papers, fusion is performed based on either the sum method or the weighted fusion method [22, 37, 39]. The significance of the work is represented in process flow presented in Algorithm 1. The algorithm automatically select the best combination of features (among 8 fusion combinations). No human intervention is required to select the combination of features (LBP, Gabor, HOG). Fusion varies from images collected at different distances and environment conditions. In this work to achieve the best rank-1 identification rate, we, first, tested 8 different fusion scenarios (see pseudo code for developed FU-GLBHG method in Algorithm 1 in Sect. 3.2.1), and finally, selected the scenario that achieved the best rank-1 identification rate for the matching scenarios investigated.

Next, we performed cross-spectral matching experiments on a demographic database. To the best of our knowledge, this is first time where cross-spectral matching of VIS and NIR images (collected in night time and at long distances) is performed, using a matching scheme based on stratification or soft biometric based clusters such as, gender, ethnicity and facial hair.

4 Experimental Results

In this section, first generated the baseline results from commercial and academic face matchers. We discussed the classification results for soft biometric traits based on proposed CNN network. Our cross-spectral face matching results under three different conditions are discussed. The first, is when we are using the LBP and LTP matchers and different distance metrics. The second, is when we utilize a set of feature descriptors (LBP, Gabor and HOG) and their fusion under multiple scenarios, including the usage of various pre-processing combinations. Finally, we investigate whether our proposed novel approach improves the baseline performance when using demographic grouping, namely a set of soft biometrics. All three conditions are discussed in detail below.

4.1 Baseline Face Matching Across Different Scenarios

We selected two different sets of face matching experiments including:

Experiment 1: Intra-spectral matching, where the NIR images (60, 90 and 120 m) in the probe set are matched against the NIR images (30 m) in the gallery set and rank-1 identification rate is generated. For this we used commercial and academic software for the baseline results.

Experiment 2: Cross-spectral matching, where the NIR images (collected in night time, at stand-off distances from 30 to 120 m) in the probe set are matched against the visible images (1.5 m) in the gallery set and rank-1 identification rate is generated. For this we used commercial and academic software for the baseline results.

Baseline Face Matchers: We used both commercial and academic FR matchers. In terms of the commercial off-the-shelf (COTS) software, we used the Identity Tools G8 (www.l1id.com) provided by L1 Systems. In the academic software, we used standard FR methods are provided by the Academic Face Identification Evaluation System (AFIES) CSU.

In the inter-spectral matching, based on the results of the baseline experiments (see in [29] for more detailed information for the experiments performed), we determined that the rank-1 identification rate is 88% for AFIES (LDA-EUC and LDA lda-soft). On the other hand, identification results are very low using COTS software, where the rank-1 identification rate is 16%. For both 90 and 120 m NIR probe images, the rank-1 identification rate is 68% and for 120 m achieves 46% from AFIES (PCA MahaCosine). With usage of demographic information, the rank-

Table 2 In this table we show the rank-scores of the *cross-spectral, cross-distance* matching scenarios for all databases used before and after the usage of demographic grouping: Experiments were run several times and the rank-1 identification rate presented here are the means values

	Cross-spectral (VIS vs. NIR), cross-distance matching (rank-1)			
VIS 1.5 m vs.	NIR 30 m	NIR 60 m	NIR 90 m	NIR 120 m
ALL DATA				
AFIES	0.19	0.22	0.21	0.15
COTS	0.48	0.03	0.02	0.01
MALE Class				
AFIES	0.34	0.33	0.26	0.18
COTS	0.52	0.03	0.02	0.02
FEMALE class				
AFIES	0.30	0.25	0.18	0.23
COTS	0.65	0.08	0.08	0.03
Asian class				
AFIES	0.37	0.39	0.32	0.28
COTS	0.70	0.48	0.04	0.05
Caucasian class				
AFIES	0.30	0.23	0.14	0.28
COTS	0.47	0.67	0.06	0.05

1 identification rate is improved from 88% to 94% for 60 m, from 68% to 79% for 90 m and 46% to 52% for 120 m distance for Male class using AFIES.

For cross-spectral matching, based on the results of the baseline experiments, we determined that the rank-1 identification rate is 48% for COTS system (VIS-NIR30 m). After the demographic information, rank-1 identification accuracy is improved from 48% to 52% for male class and to 65% for the female class, to 70% for Asian class as represented in Table 2. The rank-1 identification rate decreases with increase in the distance from 30 to 120 m from both the baseline matchers.

4.2 Demographic Grouping from Proposed CNN Architecture

In the first set of experiments, our experiments aim to illustrate how the scenario adaptable deep learning system performs for intra-spectral band, where both the training and testing data is selected from the same band (VIS-VIS) under controlled conditions. Finally, in order to determine the extent of which the performance of the classification system is affected when the standoff distance increases, we performed cross-spectral experiments, where we selected the face images from visible band (short distance) for the training and NIR face images from a distance of 30 up to 120 m away for testing. To train the CNN network two scenarios are selected (see in Sect. 2.2) including:

Scenario 1: In this total 9 databases for training and our database is used for testing.

Scenario 2: Both our databases and 9 other databases for the training and our database is used the testing (without overlap of subjects).

For intra-spectral case, for both the scenarios 1 and 2, the face images collected in the visible band (at different locations, time, expressions and illumination conditions, expressions) are selected. For the cross-spectral case, images collected in the visible band are selected to train the network and NIR images collected at a distance of 30 to 120 m are selected for the testing part. To compare the results with baseline system, we selected the bag of words model.

Classification for Gender and Ethnicity Class In our previous work [29], we reported the results for gender and ethnicity class. To perform the classification, selected LDHF database and our database in the visible band [13] for the training and our database collected at long distance for the testing. For the gender class, we used the classification results from [29] (see for more detailed information for the experiments performed) as presented in Table 3 and for the ethnicity class to improve the performance of the classification system, extended the training data with 9 databases (as discussed in Sect. 2.2) and results are presented in Table 4 for CNN and BOW model. To improve the overall classification results, combined the final results and performance reaches more than 80% for all the distances from 30 to 120 m.

Classification for Male Class into with or without Beard Categories In this CNN network is proposed for the grouping of the male class into with or without beard. Table 5, depicts the accuracy results for proposed CNN architecture with best epoch value and baseline model. For intra-spectral classification, for Scenario 1 the highest classification accuracy achieved was 53.75% from scenario 1 as presented in Table 5. For cross-spectral classification, the classification accuracy is more than 60% for scenario 1, when the visible images are selected for the training and NIR images collected at a distance of 30 m for testing. For VIS vs. NIR 60 m, the classification accuracy is more than 65% from scenario 2 and better results are achieved than the baseline model (BOW). For VIS vs. NIR 90 m, the classification accuracy is more than 60% from scenario 2 and better results are achieved than the baseline model (BOW). For VIS vs. NIR 120 m, the classification accuracy is improved from 40% (the baseline BOW model) to 67% from proposed CNN network. Based on the results, we concluded that for cross-spectral, we achieved better performance results from proposed CNN network for scenario 2. For intra-spectral, we achieved the better performance results from the baseline model (BOW) for scenario 1.

Table 3 Summary of classification results for the ethnicity and gender class based on CNN for each dataset. To perform CNN, the model is trained for the challenging testing database

	Classification accuracy		
Datasets	Gender class	Ethnicity class	
train-test	(LDHF+our) – (our)		
VIS-VIS 1.5 m	96.41	99.04	78.98
VIS-NIR 30 m	86.14	95.34	64.49
VIS-NIR 60 m	89.45	85.10	60.23
VIS-NIR 90 m	93.52	76.86	65.02
VIS-NIR 120 m	94.12	73.53	61.83

Table 4 Summary of classification results for the Ethnicity class based on CNN for each dataset. To perform CNN, model is trained on 9 different databases

Ethnicity class: scenario 1 (training with extended database)		
Datasets: train VIS (9 databases) vs. test (own database)		
Parameters	CNN	BOW
VIS 1.5 m	71.33	45.65
NIR 30 m	57.88	58.56
NIR 60 m	58.02	52.04
NIR 90 m	60.60	59.24
NIR 120 m	65.63	62.23

Classification for Female Class into Asian or Caucasian In this CNN network is proposed for the grouping of the female class into Asian or Caucasian (intra-spectral and cross-spectral classification). In Table 5, classification results are presented with best epoch value. Based on the results, we concluded that the best performance results for cross-spectral classification are achieved from proposed CNN network in comparison to the baseline BOW model. For VIS-NIR30 m, the classification accuracy reaches greater than 70% from proposed CNN network for scenario 2. The classification accuracy is improved from 59% from baseline BOW model to 71% from our proposed CNN network. For VIS-NIR 90 m, the classification accuracy is improved from 58% from baseline BOW model to 64%. The classification results are similar for VIS-NIR60 m and VIS-NIR 120 m and for intra-spectral.

Datasets: train VIS (9 databases) vs. test (own database)

4.3 *Face Matching Result Without Demographic Information Using Proposed Pre-processing Method*

Figure 6, shows the extent to which the performance of a FR system is improved by selected set of pre-processing combinations (for selected sets of combinations from *comb1*, *comb2* to *comb5*). Based on the results we concluded that the selection of a certain combination of pre-processing techniques is critical in improving cross-spectral matching performances in terms of rank-1 identification rate (see Fig. 6). Our experimental results demonstrate that when the LBP/LTP descriptors are

Table 5 Summary of classification results for the ethnicity and gender class based on CNN for each dataset. To perform CNN, the model is trained for the challenging testing database

Classification accuracy for male class into with or without beard					
Datasets: train vs. test					
VIS vs.	VIS 1.5 m	NIR 30 m	NIR 60 m	NIR 90 m	NIR 120 m
Scenario 1: CNN	53.75	60.36	63.21	61.25	62.50
Scenario 1: BOW	69.46	62.14	62.68	61.607	40.36
Scenario 2: CNN	52.64	55.53	68.51	63.94	67.31
Scenario 2: BOW	45.53	40.00	38.04	37.32	37.67
Classification accuracy for female class into Asian or Caucasian class					
Scenario 1: CNN	74.24	66.67	57.95	57.58	64.01
Scenario 1: BOW	75.75	56.06	55.68	56.06	57.58
Scenario 2: CNN	71.12	71.00	68.50	64.00	60.00
Scenario 2: BOW	36.00	59.50	66.00	58.00	62.00

employed, better results are obtained for *comb2*, when first we use *DN* and then *PN* (single scale self quotient). This order (*comb 2*) results in 36% rank-1 identification rate (e.g. for LBP-CHI) and similar for other descriptors and distance metrics (see Fig. 6), a 25% improvement in comparison to when using the raw images without selected pre-processing combination (No Norm).

Our experimental results indicated that for LBP/LTP based methods, the highest overall accuracy i.e. 36% and, hence, these results are not satisfying. Thus, we investigated an alternative, more complicated, but much more efficient approach. We first, replaced the DT and CHI distance transform methods into a kernel subspace method [37].

To train the system, we used 60% of the data, 63 subjects and 8 samples per subject (4 face images in VIS and 4 face images in NIR band for each subject). The rest of the data is used for testing, i.e. 4 face images for each test subject in the visible band for gallery set and 4 face images for the NIR probe set. Note that there is no subject overlap between the training and test sets.

We followed the same procedure as applied before when using the LBP and LTP descriptors including: usage of various pre-processing combinations based on proposed image restoration approach following *PN* method. We investigated 5 different combinations for pre-processing and, finally, selected the best combination based on the performance results. We concluded that (as shown in Fig. 7), the selection of the combination of pre-processing technique is critical in improving the matching performance for each selected feature descriptor.

First, we selected the PN, for each feature descriptor and identified the best combination that provided us with best rank-1 identification rate as shown in Fig. 7. Our experimental results demonstrate that when HOG/KLDA descriptor is employed, better results are obtained for *comb2*, i.e. when first we use denoising and then isotropic smoothing. This ordered pre-processing of the face images results in 65% rank-1 identification rate (see bottom-left table in Fig. 7). When the LBP/KLDA descriptor is used, better results are obtained for *comb5*, when, first, we use super-resolution, second, we use denoising and, then, single scale self quotient normalization. This ordered pre-processing of the face images results in 53% rank-1 identification rate (see top-left Fig. 7). Finally, when the Gabor/KLDA descriptor is used, better results are obtained for *comb1*, i.e. when we use the Tann and Triggs photometric normalization technique. This order results in 58% rank-1 identification rate (see top-right Fig. 7). The selection of pre-processing combination and PN techniques depends on the feature descriptor used.

4.4 Face Matching Results Without Demographic Information Using Proposed Fusion Scheme

As we discussed above, we compare visible to NIR face images under diverse conditions. In the previous subsection we empirically determined which feature descriptor and pre-processing combination results in the best rank-1 identification rate using proposed FU-GLBHG method. We investigated 8 different fusion scenarios based on sum and weighted fusion as presented in pseudo code Algorithm 1. Here, schemes $A1$, $A2$ and $A3$ are without any fusion and based on individual methods. For sum fusion method, the fusion of normalized distance scores, $z1$ for Gabor/KLDA, $z2$ for LBP/KLDA and $z3$ for HOG/KLDA is performed using four different cases (presented in pseudo code Algorithm 1). When the weighted fusion method is used, the weights are assigned to the normalized scores based on the rank-1 identification rates that were achieved from the other independent methods tested. When the weighted fusion scheme is used, weights are automatically selected for each of combination (*comb1 to comb5*). The fusion of normalized distance scores with assigned weights is performed using four different cases (Gabor-HOG, Gabor-LBP, LBP-HOG and Gabor-LBP-HOG). Fusion scenarios are selected from $A4$ to $A11$ as presented in pseudo code Algorithm 1. The first four fusion schemes are based on sum fusion ($A4$ to $A7$) and rest on weighted fusion ($A8$ to $A11$).

The accuracy of the system is represented as rank-1 identification rate ($R1$) for all the five pre-processing combinations (i.e. for *comb1*, $A1$ is 58% (only Gabor), $A2$ is 45% (only LBP) and $A3$ is 60% (only HOG) for without fusion). The best performance result is 69% and achieved for *comb3* from sum fusion scheme $A7$ (z1+z2+z3) based on face recognition test. For rest of the four combinations, best performance scores from *comb1* to *comb5*, are achieved from weighted fusion scheme for $Gabor$, LBP and HOG. The rank-1 scores are close for *comb1* and *comb2*. Finally, we selected the *comb1* for All-Database set (for probe images at 30 m distance in NIR band) (Table 6).

Algorithm 1 Proposed FU-GLBHG method

```
procedure SELECTED SCENARIO
    z1 ← Gabor Features
    z2 ← LBP Features
    z3 ← HOG Features
    w1 = 0.60 and w2 = 0.40 ← weight1 and weight2 for 2 set of features
    w1 = 0.70, w2 = 0.15 and w3 = 0.15 ← weight1, weight2 and weight 3 for 3 set of features
    i ← Selected Scenario
    A(i) ← Accuracy of System or Rank-1 Identification Rate from Selected Scenario
top:
A(1)=z1, A(2)=z2 and A(3)=z3
Based on Sum Fusion scenario (2 features)
A(4)=z1+z2
A(5)=z1+z3
A(6)=z2+z3
Based on Sum Fusion scenario (3 features)
A(7)=z1+z2+z3
Based on Weighted Fusion scenario (2 features)
    if A(1) > A(2) then
A(8)=w1*z1+w2*z2
    else
A(8)=w1*z2+w2*z1
    if A(1) > A(3) then
A(9)=w1*z1+w2*z3
    else
A(9)=w1*z3+w2*z1
    if A(2) > A(3) then
A(10)=w1*z2+w2*z3
    else
A(10)=w1*z3+w2*z2
Based on Weighted Fusion scenario (3 features)
    if A(1) > A(2) AND A(1) > A(3) then
A(11)=w1*z1+w2*z2+w3*z3
    else if A(2) > A(1) AND A(2) > A(3) then
A(11)=w1*z2+w2*z1+w3*z3
    else if A(3) > A(1) AND A(3) > A(2) then
A(11)=w1*z3+w2*z1+w3*z2
    Accuracy ← highest rank-1 identification rate from i : 1 to 11 selected Scenarios Accuracy=maxval(A(i))
```

4.5 *Experimental Results with Usage of Demographic Information Using Proposed Pre-processing and Fusion Scheme*

The use of demographic information (soft biometrics) is proposed to improve the performance of traditional FR systems. Soft biometric traits are physical and behavioral features (weight, height, gender, ethnicity etc.), and offer several advantages over the traditional systems. In our work, we considered three main strata or traits: gender, ethnicity and facial hair. For female data, grouped the database into Female Asian (FA) and Female Caucasian (FC) and for male into male with beard (MWB) and male without beard (MWOB) strata. The selection of grouping for female class into FA and FC, for male class into MWB and MWOB is performed based on identification results from face recognition system. We conducted face matching experiments for female and male class into Asian and

Table 6 Cross-spectral face matching results (Gallery in VIS and Probe images in NIR band): Rank-1 identification rate for All-Database, for selected combinations of our proposed *pre-processing techniques. comb1, comb2, comb3, comb4* and *comb5*. Accuracy of system (R1: rank-1 identification rate) based on distance scores: Gabor/KLDA: A1, LBP/KLDA: A2 and HOG/KLDA: A3. Normalized distance scores are z1: Gabor/KLDA, z2: LBP/KLDA and z3: HOG/KLDA

Pre-processing combination *comb*	1	2	3	4	5
Rank-1 identification rate	R1	R1	R1	R1	R1
A1	0.58	0.55	0.55	0.52	0.53
A2	0.45	0.50	0.48	0.48	0.53
A3	0.60	0.65	0.55	0.62	0.60
A4	0.68	0.66	0.63	0.61	0.70
A5	0.65	0.60	0.61	0.57	0.65
A6	0.60	0.61	0.60	0.66	0.60
A7	0.69	0.70	0.69	0.63	0.66
A8	0.73	0.68	0.60	0.61	0.71
A9	0.67	0.62	0.63	0.59	0.65
A10	0.64	0.62	0.65	0.67	0.59
A11	0.69	0.73	0.62	0.67	0.66

Caucasian class. For male class into Asian and Caucasian the maximum rank-1 identification accuracy reaches to 70%. On the other hand, for male class with and without beard, we achieved better results and achieved more than 80% rank-1 identification accuracy.

Based on the results using face recognition system, for female stratum as shown in Fig. 8, we concluded that selection of combination of pre-processing technique *(comb1, comb 2, ..., comb5)* is the key factor in improving the matching performance. Our experimental results demonstrate that for FA stratum, better results are obtained for *comb5*, when first we use *SR*, second we use *DN* and then *PN* (adaptive single scale retinex named as norm 2 see in top-right Fig. 8). The performance of the system increases (rank-1 identification rate) from 55% to 75%. For FC stratum, results are similar for all the combinations (see in bottom-left). We selected *comb5*, based on our results for first 5 ranks and this combination results in 53% rank-1 identification rate. For a comparison between with and without the usage of stratification of the database results are presented in bottom-right.

Same set of experiments is repeated for male strata into male with and without beard. Based on the face matching results, we evaluated that *comb5* for MWOB and *comb2* for MWB are selected as the best combinations. The rank-1 identification rate with best results using the five pre-processing combinations for MWB and MWOB are shown in Fig. 9. Based on the results, we concluded that better performance results are obtained from *comb2* for MWB and from *comb5* for MWOB. The performance of the system increases (rank-1 identification rate) from 58% to 75% for MWOB (bottom left), for *comb5*, 63% to 82% for MWB (top right).

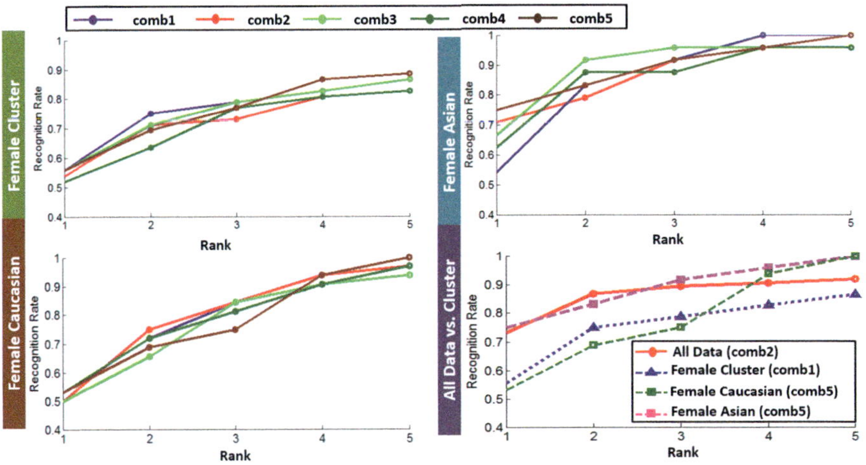

Fig. 8 Cross-spectral matching results for **Female Stratum**: CMC curves comparing the performance for the face images restored from our proposed method with raw images

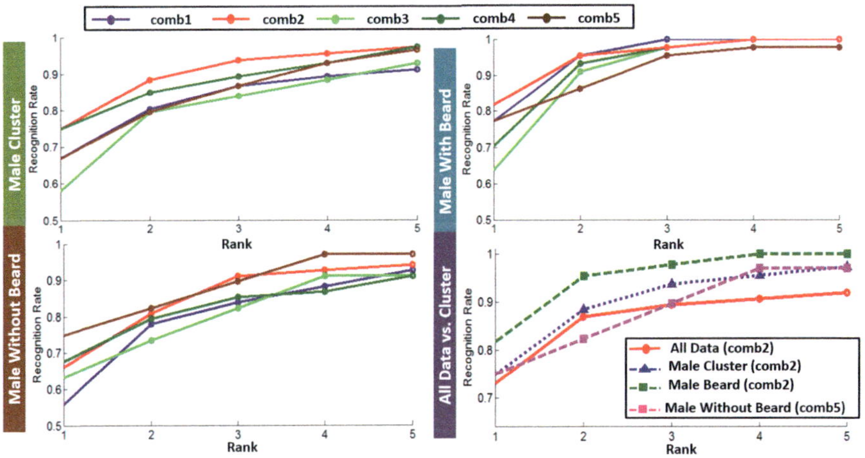

Fig. 9 Cross-spectral matching results for **Male Cluster**: CMC curves comparing the performance of FR system for the face images restored from our proposed method with raw images

4.5.1 Cross-Matching Results for Long Distances

We repeated each experiment (five scenarios: All Database, MWB, MWOB, FA and FC) 10 times, where we randomly selected the training and testing data and performed the feature extraction and finally system evaluation to remove the bias in the results. For the scenario without the usage of demographic information database (All Data), the best rank based identification accuracy results (rank-1 to rank-5) are reported in Table 7. For the selected set of strata, the best rank based identification accuracy results are reported in Table 8.

We performed the same set of experiments (10 times) for NIR database at distance of 60, 90 and 120 m for the five scenarios (All Database, FA, FC, MWB and MWOB).

Based on the results for the scenario with all the database for 4 different distances NIR 30, 60, 90 and 120 m (see Table 7), it was determined that for all distances, better performance results are achieved from our proposed methods in comparison to the commercial off the shelf (COTS) G8 system (L1 Systems). For 30 m distance, performance reaches almost 48% for COTS system, whereas, the accuracy reaches almost 73% when using our proposed approach. For the 30 m distance dataset, the rank-1 identification accuracy results are very close to each other for *comb1* and *comb2*. We have selected *comb1* based on a comparison among the first 5 rank-1 identification rate.

For the 60 m NIR images collected at long standoff distance, the rank-1 identification rate is 10% when the COTS system software is used. We achieved better performance results from our proposed pre-processing and fusion schemes, i.e. the rank-1 identification rate in our approach is more than 5 times improved, or 56% for *comb2* (see Table 7). For the 90 m images collected at long standoff distance, the rank-1 identification rate is 1% when the COTS system software is used. We achieved better performance results from our proposed pre-processing and fusion schemes, i.e. the rank-1 identification rate in our approach is more than 30 times improved, or 30% for *comb3*. Finally, for 120 m images the rank-1 identification rate in our approach is more than 28 times improved, or 28% for *comb3*.

Without usage of demographic information (All Database), the rank-1 identification rate is improved from 48% for COTS system, 19% for AFIES to 73% when using our proposed approach (see in Table 9) 30 m distance. For 60 m distance, the rank-1 identification rate is improved from 3% for COTS system, 22% for AFIES to 56% using proposed approach. There is an increase in rank-1 identification rate for 90 and 120 m distance as represented in Table 9.

The identification experiments were performed for the database with demographic information or soft biometric traits (FA, FC, MWB and MWOB) and each experiment is repeated 10 times (randomly selected training and test data) and results from best set is reported in Table 8. For 30 m distance, without stratification for all the database performance results achieved were 73% from our proposed method (see in Table 7) and after stratification it was determined that the better performance results are achieved, for FA 75%, for MWOB 75% and for MWB 82%. For 90 and 120 m distance there is great improvement in the results for the data with demographic information. For 90 m, the performance of the systems increases from 30% to 41% for FA, 43% for MWB and 38% for MWOB. For 120 m, the performance of the system increases from 28% to 38% for FA, and 38% for MWB.

The overall best results are attained for FA, MWB and MWOB (soft biometric traits) for 30, 90 and 120 m in comparison to all the database (without stratification) as illustrated in Table 8. For one of soft biometric trait, FC, we did not achieve as good results as originally expected. A valid reason for these results is the type of

Table 7 Cross-spectral Matching Results (Gallery in VIS (1 m-indoor) under controlled conditions and Probe images in NIR): Rank Identification Rate for probe images at 4 different distances in nighttime environment: 30, 60, 90 and 120 m distance

Rank identification accuracy	R1	R2	R3	R4	R5
Probe images at 30 m					
Pre-processing combination					
comb1	**0.73**	**0.86**	**0.88**	**0.91**	**0.92**
comb2	0.73	0.86	0.89	0.90	0.91
comb3	0.69	0.83	0.85	0.91	0.94
comb4	0.67	0.83	0.89	0.91	0.92
comb5	0.70	0.78	0.82	0.85	0.91
COTS	**0.48**	**0.51**	**0.55**	**0.60**	**0.62**
Probe images at 60 m					
comb1	0.44	0.59	0.69	0.74	0.79
comb2	**0.56**	**0.67**	**0.76**	**0.84**	**0.86**
comb3	0.51	0.61	0.70	0.77	0.85
comb4	0.48	0.63	0.73	0.77	0.83
comb5	0.39	0.54	0.61	0.68	0.78
COTS	**0.01**	**0.02**	**0.03**	**0.04**	**0.05**
Probe images at 90 m					
comb1	0.20	0.30	0.40	0.45	0.52
comb2	0.25	0.33	0.40	0.47	0.53
comb3	**0.30**	**0.39**	**0.49**	**0.54**	**0.59**
comb4	0.22	0.30	0.36	0.40	0.44
comb5	0.24	0.30	0.37	0.41	0.47
COTS	**0.01**	**0.01**	**0.02**	**0.02**	**0.04**
Probe images at 120 m					
comb1	0.16	0.26	0.37	0.41	0.47
comb2	0.23	0.37	0.44	0.51	0.54
comb3	**0.28**	**0.40**	**0.48**	**0.52**	**0.58**
comb4	0.20	0.30	0.36	0.42	0.46
comb5	0.18	0.28	0.32	0.36	0.43
COTS	**0.01**	**0.01**	**0.02**	**0.04**	**0.04**

stratum and very small difference in the subjects' appearance. For the 60 m distance, the results are very close, with and without using stratification. Our proposed methods including: image restoration, feature selection, decision level fusion and finally the selection of soft biometric traits (stratification of database) result in significant improvement in face identification results than available commercial FR matcher.

Table 8 Cross-spectral matching results for Soft Biometric Traits (Gallery in VIS (1 m-indoor) under controlled conditions and Probe images in NIR): Rank identification accuracy for probe images at 4 different distances in nighttime environment: 30, 60, 90 and 120 m distance

Rank identification accuracy	R1	R2	R3	R4	R5
Female Asian:					
Distances					
30 m	0.75	0.83	0.91	0.95	1
60 m	0.37	0.58	0.79	0.96	0.96
90 m	0.41	0.54	0.67	0.87	1
120 m	0.38	0.58	0.75	0.91	0.96
Female Caucasian:					
30 m	0.53	0.68	0.75	0.93	1
60 m	0.50	0.66	0.72	0.84	0.84
90 m	0.28	0.47	0.63	0.78	0.88
120 m	0.25	0.47	0.56	0.65	0.75
Male with beard:					
30 m	0.82	0.95	0.97	1	1
60 m	0.52	0.73	0.89	0.98	0.98
90 m	0.43	0.54	0.75	0.84	0.90
120 m	0.38	0.52	0.68	0.80	0.82
Male without beard:					
30 m	0.75	0.82	0.89	0.97	0.97
60 m	0.49	0.72	0.80	0.84	0.88
90 m	0.38	0.53	0.58	0.65	0.71
120 m	0.27	0.43	0.50	0.61	0.65

5 Conclusion

In this work, we study the challenges of matching NIR face images collected in heterogeneous environments i.e. when the face images are captured in night time, at variable long standoff distances against the visible (good quality) images. We investigated the impact of usage of demographic information in terms of ethnicity, gender and facial hair to the performance of cross-spectral face recognition system. We proposed a deep convolutional neural network based architecture to classify the visible and multi-distance NIR face images for the male class into with or without beard and female class into Asian and Caucasian class. The proposed network is designed to make it adaptable with intra-spectral (VIS vs. VIS) and cross-spectral (VIS vs. NIR) scenarios at variables distances. In the experiments we performed, we trained the model when using a multi-band database, where the training data is selected from the visible band dataset (controlled conditions at a short standoff distance) and the testing data is selected from the NIR band multi-distance face images (30 up to 120 m).

Table 9 In this table we investigate the *cross-spectral, cross-distance* matching scenarios for database before and after the usage of proposed image restoration and demographic grouping of male class into with or without beard and female class into Asian and Caucasian class: Experiments were run several times and the rank-1 identification accuracy presented here are the means. Proposed-Set 1: Only Proposed Fusion Scheme applied to the images, Proposed-Set 2: Both proposed image restoration and fusion Scheme applied to the images. For Set 2 with Male and Female Class further sub-clustering is performed: MWB, MWOB, FA and FC class

VIS 1.5 m vs.	Cross-spectral (VIS vs. NIR), cross-distance matching (rank-1)			
	NIR 30 m	NIR 60 m	NIR 90 m	NIR 120 m
ALL DATA				
AFIES	0.19	0.22	0.21	0.15
COTS	0.48	0.03	0.02	0.01
Proposed-set 1	0.56	0.34	0.21	0.16
Proposed-set 2	**0.73**	**0.56**	**0.30**	**0.28**
MALE class				
AFIES	0.34	0.33	0.26	0.18
COTS	0.52	0.03	0.02	0.02
Proposed-set 1	0.34	0.33	0.26	0.18
Proposed-set 2: MWB	**0.82**	0.52	**0.43**	0.38
Proposed-set 2: MWOB	**0.75**	0.49	**0.38**	0.27
Female class				
AFIES	0.30	0.25	0.18	0.23
COTS	0.65	0.08	0.08	0.03
Proposed-set 1	0.37	0.39	0.32	0.28
Proposed-set 2: FA	**0.75**	0.37	0.41	0.38
Proposed-set 2: FC	0.53	0.50	0.28	0.25

Based on experiments, the proposed CNN architecture provided us with significant classification results for the selected challenging databases. The experimental results show that from the proposed CNN network, we achieved significant improvement in the classification results when compared to the baseline bag of words model. For the female class into Asian and Caucasian, for cross-spectral scenario, e.g. for VIS-NIR120 m, the classification is improved from 59% to 71% for VIS vs. NIR 60 m and 58% to 64% for VIS vs. NIR 120 m.

We proposed a pre-processing method, composed of a set of image restoration approach and *PN* techniques, as well as our proposed scenario dependent fusion schemes were evaluated under 5 different matching scenarios, (i) Original Database and the following sub-sets (ii) Female Asian, (iii) Female Caucasian, (iv) Male With Beard and (v) Male Without Beard. Based on our experimental results, when using stratification of the datasets resulted in significant improvement in our rank-1 identification rate for each cross-spectral scenario investigated. For long distances particularly at 90 and 120 m, there is significant improvement in rank identification rate (from rank 1 to rank 5). The best results we achieved for FA, MWB and MWOB. For FC, the results are not satisfactory and one of the main reason is, less between class variation e.g. eyes color, hair color etc. On the contrary, for FA, MWB and MWOB, involves subjects from different races with different eyes color, hair color and size of the faces. These are the factors responsible for better performance. Since, the database was very limited for African American class in comparison to Caucasian and Asian class and it belongs to male class. To address this challenge, we selected only Caucasian and Asian for female class and for male class selected most discriminating features (with and without beard). In future, we will collect more database to consider other groups for ethnicity classification.

Based on the experimental results, we conclude that: First, a CNN can be used to classify the data in terms of ethnicity and gender class into with or without beard when using both constrained and unconstrained face datasets. We show that the proposed image restoration approach and fusion schemes achieve significantly better performance across all the scenarios compared to the commercial FR matcher. We evaluated that for long distances particularly for 60, 90 and 120 m, our proposed system can be utilized to improve cross-spectral matching performance on diverse scenarios when compared to the commercial FR matcher. In future work, we plan to investigate other factors, for example generating subsets based on facial expression, pose etc. in our experiments.

Acknowledgements This work was sponsored in part through a grant from the Office of Naval Research (ONR), contract N00014-09-C-0495, "Distribution A—Approved or Unlimited Distribution". The authors are also grateful to WVU faculty and students from West Virginia University, especially, Dr. J. Dawson and John Dollen, who assisted us with parts of this work. The views expressed in this manuscript are those of the authors and do not reflect the official policy or position of the Department of Defense, or the U.S. Government.

References

1. Li SZ, Chu RF, Liao SC, Zhang L (2007) Illumination invariant face recognition using near-infrared images. IEEE Trans Pattern Anal Mach Intell 29(4):627–639
2. Bourlai T, Dollen J, Mavridis N, Kolanko C (2012) Evaluating the efficiency of a night-time, middle-range infrared sensor for applications in human detection and recognition. In: Proceedings of SPIE defense, security, and sensing, Baltimore, May 2012, pp 83551B–83551B-12
3. Ghiass RS, Arandjelović O, Bendada A, Maldague X (2014) Infrared face recognition: a comprehensive review of methodologies and databases. Pattern Recogn 47(9):2807–2824
4. Mansanet J, Albiol A, Paredes R (2016) Local deep neural networks for gender recognition. Pattern Recogn Lett 70:80–86
5. Jain AK, Dass SC, Nandakumar K (2004) Can soft biometric traits assist user recognition? In: Proceedings of SPIE, vol 5404, pp 561–572
6. Nixon MS, Correia PL, Nasrollahi K, Moeslund TB, Hadid A, Tistarelli M (2015) On soft biometrics. Pattern Recogn Lett 68:218–230
7. Mery D, Bowyer K (2015) Automatic facial attribute analysis via adaptive sparse representation of random patches. Pattern Recogn Lett 68:260–269
8. Park U, Jain AK (2010) Face matching and retrieval using soft biometrics. IEEE Trans Inf Forensics Secur 5(3):406–415

9. Bourlai T, Mavridis N, Narang N (2016) On designing practical long range near infrared-based face recognition systems. Image Vis Comput 52:25–41
10. Liu M, Xie W, Chen X, Ma Y, Guo Y, Meng J, Yuan Z, Qin Q (2011) Heterogeneous face biometrics based on Guassian weights and invariant features synthesis. In: Proceedings of IEEE 2nd international conference on computing, control and industrial engineering (CCIE), vol II, Aug 2011, pp 374–377
11. Chen J, Yi D, Yang J, Zhao G, Li SZ, Pietikainen M (2009) Learning mappings for face synthesis from near infrared to visual light images. In: Proceedings of IEEE conference on computer vision and pattern recognition (CVPR), pp 156–163
12. Klare B, Jain AK (2010) Heterogeneous face recognition: matching NIR to visible light images. In: Proceedings of 20th international conference on pattern recognition (ICPR), Washington, DC, 2010, pp 1513–1516
13. Maeng H, Choi H-C, Park U, Lee S-W, Jain AK (2011) Nfrad: near-infrared face recognition at a distance. In: Proceedings of international joint conference on biometrics (IJCB). IEEE, pp 1–7
14. Maeng H, Liao S, Kang D, Lee S-W, Jain AK (2013) Nighttime face recognition at long distance: cross-distance and cross-spectral matching. In: Proceedings of computer vision–ACCV'12, Berlin/Heidelberg, vol II, pp 708–721
15. Yi D, Liu R, Chu R, Lei Z, Li SZ (2007) Face matching between near infrared and visible light images. In: International conference on biometrics, Springer, Berlin, Heidelberg, Aug 2007, pp 523–530
16. Osia N, Bourlai T, Hornak L (2018) Facial Surveillance and Recognition in the Passive Infrared Bands. Surveillance in Action, Springer, Cham, pp 127–145
17. Osia N, Bourlai T (2017) Bridging the spectral gap using image synthesis: a study on matching visible to passive infrared face images. Mach Vis Appl 28(5–6):649–663
18. Wang R, Yang J, Yi D, Li SZ (2009) An analysis-by-synthesis method for heterogeneous face biometrics. In: International conference on biometrics, Springer, Berlin, Heidelberg, Jun 2009, pp 319–326
19. Zhang Z, Wang Y, Zhang Z (2011) Face synthesis from near-infrared to visual light via sparse representation. In: Proceedings of IEEE international joint conference on biometrics (IJCB), Washington, DC, Oct 2011, pp 1–6
20. Goswami D, Chan CH, Windridge D, Kittler J (2011) Evaluation of face recognition system in heterogeneous environments (visible vs NIR). In: Proceedings of IEEE international conference on computer vision (ICCV) workshops, Barcelona, vol 5558, Nov 2011, pp 2160–2167
21. Raghavendra R, Dorizzi B, Rao A, Hemantha Kumar G (2011) Particle swarm optimization based fusion of near infrared and visible images for improved face verification. Pattern Recogn 44(2):401–411
22. Omri F, Foufou S, Abidi M (2013) Pixel level fusion of multispectral face images: short review. In: Proceedings of 7th IEEE Gulf cooperation council (GCC) conference and exhibition, Doha Qatar, Nov 2013, pp 595–600
23. Omri F, Foufou S, Abidi M (2014) NIR and visible image fusion for improving face recognition at long distance. In: International conference on image and signal processing, Springer, Cham, Cherbourg, Normandy, France, pp 549–557
24. Kang D, Han H, Jain AK, Lee SW (2014) Nighttime face recognition at large standoff: cross-distance and cross-spectral matching. Pattern Recogn 47(12):3750–3766
25. Narang N, Bourlai T (2016) On the effectiveness of statistical hypothesis testing in infrared-based face recognition in heterogeneous environments. In: 2016 IEEE/ACM international conference on advances in social networks analysis and mining (ASONAM), Aug 2016, pp 1278–1285
26. Vedaldi A, Lenc K (2015) MatConvNet – convolutional neural networks for MATLAB. In: Proceedings of the 23rd ACM international conference on multimedia, New York, pp 689–692

27. Levi G, Hassner T (2015) Age and gender classification using convolutional neural networks. In: IEEE conference on computer vision and pattern recognition (CVPR) workshops, June 2015
28. Parkhi OM, Vedaldi A, Zisserman A (2015) Deep face recognition. Proc Br Mach Vis 1(3):6
29. Narang N, Bourlai T (2016) Gender and ethnicity classification using deep learning in heterogeneous face recognition. In: 9th IAPR international conference on biometrics ICB, Halmstad, June 2016
30. Whitelam C, Jafri Z, Bourlai T (2010) Multispectral eye detection: a preliminary study. In: 2010 20th international conference on pattern recognition (ICPR). IEEE, pp 209–212
31. Nicolo F, Schmid NA (2012) Long range cross-spectral face recognition: matching SWIR against visible light images. IEEE Trans Inf Forensics Secur 7(6):1717–1726
32. Thomaz CE, Giraldi GA (2010) A new ranking method for principal components analysis and its application to face image analysis. Image Vis Comput 28(6):902–913
33. Hond D, Spacek L (1997) Distinctive descriptions for face processing. In: BMVC, number 0.2, pp 0–4
34. Gilad F, Raanan F (2011) Image and video upscaling from local self examples. ACM Trans Graph (TOG) 30(2):1–11
35. Bourlai T, Ross A, Jain A (2009) On matching digital face images against scanned passport photos. In: Proceedings of first IEEE international conference on biometrics, identity and security (BIDS), Tampa, Sept 2009
36. Struc V (2012) The inface toolbox for illumination invariant face recognition, Feb 2012
37. Tan X, Triggs B (2010) Enhanced local texture feature sets for face recognition under difficult lighting conditions. IEEE Trans Image Process 19(6):1635–1650
38. Dalal N, Triggs B (2005) Histograms of oriented gradients for human detection. In: Proceedings of international conference on computer vision & pattern recognition (CVPR), vol 2, June 2005, pp 886–893
39. Narang N, Bourlai T (2015) Face recognition in the SWIR band when using single sensor multi-wavelength imaging systems. Image Vis Comput 33(0):26–43

Quality and Match Performance Analysis of Band-Filtered Visible RGB Images

Jeremy Dawson, John Goodwyn, S. Means, and Jason Crakes

Abstract Face recognition performance in operational scenarios is can be improved by using cameras that capture multispectral or hyperspectral images at specific bands within the visible spectrum. Band-selected images have shown promise to improve face recognition performance, but the requisite camera systems needed to achieve multi-filter or hyperspectral imaging are often to complex and cost-prohibitive for many law enforcement applications. In order to find a more cost-effective solution, the work presented here aims to determine if simple band-filtered images, captured by placing bandpass filters on conventional RGB imagers, show any application advantages over broad-spectrum visible facial imagery. After data collection was completed, matching studies were performed to determine what performance enhancement, if any, is gained using band-filtered imaging. Results indicate that image quality may play a bigger role in the facial recognition performance of band-filtered images rather than simple band-filtering alone, warranting further study in this area.

Keywords Facial recognition · Unconstrained · Band selection · Multi-spectral

1 Introduction

Operational surveillance scenarios often present challenges to facial recognition systems, including uncontrolled lighting conditions and non-optimal pose angles. There are many scenarios in which facial recognition performance should be improved to enhance field operations, specifically: fixed location surveillance operations and end-to-end video analytics. One method of improving face recognition performance in operational scenarios is to employ cameras that capture multispectral or

J. Dawson (✉) · J. Goodwyn · S. Means · J. Crakes
Lane Department of Computer Science and Electrical Engineering, West Virginia University, Morgantown, WV, USA
e-mail: jeremy.dawson@mail.wvu.edu

T. Bourlai et al. (eds.), *Securing Social Identity in Mobile Platforms*, Advanced Sciences and Technologies for Security Applications,
https://doi.org/10.1007/978-3-030-39489-9_6

hyperspectral images at specific bands within the visible spectrum. These band-selected images have shown promise to improve face recognition performance under operational conditions. However, multi-filter or hyperspectral approaches are often to complex and cost-prohibitive for many law enforcement applications. In order to find a more cost-effective solution, the work presented here aims to determine if simple band-filtered images, captured by placing bandpass filters on conventional RGB imagers, show any application advantages over broad-spectrum visible facial imagery. For this work, the specific performance advantages evaluated were: (1) facial recognition matching performance using commercial facial matchers and (2) image quality assessed using machine learning techniques. The flowing section will provide a review of the relevant literature pertaining to multispectral and hyperspectral imagery and their application in the field of human facial recognition.

1.1 Literature Review

1.1.1 Image Quality Assessment

Image Quality Assessment is an essential part of the creation of a face recognition system. In [1] Wang et al. discuss the difficulties of image quality assessment. The way the human eye perceives quality is the most accurate, but the methods employed in image quality systems to mimic this are both computationally expensive and monetarily expensive to include. There are also error sensitivity based frameworks, but these are highly reliant on many assumptions, including having a reference image of perfect quality, which is very difficult to achieve. They go on to propose a new method, which models image degradation as structural distortions instead of errors, which has shown promising results.

Wang and Bovik [2] demonstrate two different approaches to image quality assessment, the mathematical approach and measurement methods that consider human visual system characteristics. They propose a new universal quality index which is modeled as a combination of three factors: loss of correlation, luminance distortion, and contrast distortion. Although their approach is highly mathematical, and no human visual system model is used, it has shown to perform significantly better than other widely used models.

Wang et al. [3] further image quality research by discussing traditional methods that attempt to quantify the differences between a distorted image and a reference image using a variety of known properties of the human visual system. The authors propose an alternative complementary framework based on the degradation of structural information. The new system performed well compared to other methods across a database containing images of various quality.

Sellahewa and Jassim [4] discuss how automated face recognition systems are affected by lighting condition changes between enrollment and identification images and how it leads to a decrease in recognition accuracy. They propose a quality-based approach to the face recognition problem by measuring the illumination quality, using a global quality normalization scheme, and using the illumination quality measure to adaptively select weighting parameters. Our image quality assessment is similar in nature but focused on finding the best image and wavelength for hyperspectral face recognition.

Likewise, Abaza et al. [5] discussed the importance of image quality in automated face recognition systems. They proposed an effective image quality index in order to provide real-time feedback for reducing the number of poor-quality face images acquired during enrollment and validation in a face recognition system. The authors use a database of visible face images and find values for several image quality metrics in which they discuss which quality metric is the most important in face recognition and propose a new face image quality index that combines these metrics into one score. This process is an outline of how we have set up our system, but we have optimized the metrics used for hyperspectral face images and the process of determining which wavelength best suits hyperspectral face recognition.

The topic of image quality in the visible spectrum been heavily discussed in the open literature, however, the exploration of image quality in other imaging spectrums is still largely an unexplored topic. Martin and Bourlai [6] study tattoo images in biometrics and forensic-related applications. In this paper, they discuss detection in the shortwave infrared spectrum and use image quality metrics to determine the best wavelength range for the tattoo modality and how different skin pigments affect the outcome. This paper helps us to realize that image quality can vary based on the wavelength the image is captured in. We took this into account when creating our system and employ similar methods to determine the best wavelength for our work.

1.1.2 Face Image Quality Assessment

Best-Rowden et al. [7] define biometric sample quality as a measure of a sample's utility to automatic matching. It should be an indicator of recognition performance; wherein poor quality biometric samples cause recognition to fail and good quality samples will pass. They discuss the benefits of being able to detect poor quality faces for multiple reasons, including being able to deal with them properly before entering them into a recognition system. They cite this to be of importance, especially in a security scenario, where a subject might want to evade camera systems by obstructing faces, moving to cause blur, or various other reasons. For these reasons, face quality metrics are important to put into place, and include pose, illumination, expression, occlusion, resolution, and other intrinsic or extrinsic face properties. In their study, the compare face quality assessment by human quality ratings (matcher-independent) and quality values computed from a similarity scores obtained from face matchers (matcher-dependent). The results of their experiments showed that automatic systems rejected the bottom 5% of images, while humans rejected the bottom 13%. This shows that automatic systems are promising, and they suggest that in the future face quality systems may improve with the implementation of face detection and alignment prior to assessing the quality.

Khryashchev et al. [8] discuss the challenges of face recognition in both indoor and outdoor video surveillance applications. The discuss the various challenges surrounding real time face detection, including face image quality, uncooperative subjects, and uncontrolled environments. The face quality metrics they address are motion blur, illumination, small face regions, face rotation, compression artifacts, and focus. To combat these issues, the authors proposed an automatic face quality assessment in which images are fed into a face detection algorithm, are post-processed, and their quality is evaluated. They then discard low quality images and only images of a suitable quality are used in further analysis. From their experiments, they determined that using quality measures in their application increased the recognition accuracy and also reduced the computational complexity of the system. In [9] the authors go on to discuss the use of image quality systems as a strategy to choose face images with the best quality from a group of images taken from video (i.e. surveillance cameras) to improve real-time recognition systems. The authors come to the conclusion that in low light scenarios, the blur measure is better correlated to recognition systems, whereas in normal and high illumination scenarios their proposed measure based on symmetry of landmark points outperforms all other quality measures tested.

Chen et al. [10] also address the topic of face image quality and its role in face recognition in security applications. Their application first uses a Convolutional Neural Network to normalize faces in a frame, this is typically very slow process causes degradation of the face since most systems have not perfected this method, but they have adapted this process to be more efficient using a circle around the face and using its properties to calculate the best rotation. They then take a Deep Learning approach to face quality by creating a feature vector and finding the value of rank weight. Their method assumes that images of the same database are all a similar quality. This approach does not always work, especially in security applications, such as mugshot images, where image quality can vary depending on the police department that collected the images.

1.1.3 Hyperspectral Face Detection and Recognition

Di et al. [11] discuss face recognition studies completed using hyperspectral imagery in the visible spectrum. The authors focus on absorption bands related to hemoglobin, citing them as being more discriminative than other bands and therefore selecting feature bands based on the physical absorption characteristics. In their experiments, they found that this selection outperformed using conventional RGB color bands, single band selections, and using the whole band.

Pan et al. [12] discuss human face recognition using hyperspectral imagery in the near infrared range using the subsurface tissue structure which varies greatly from person to person, but stays relatively stable over time. They propose a system that exploits spectral measurements for multiple facial tissue types and show that it performs well over time in the presence of changes in facial pose and expression. In [13] they discuss the applications that this algorithm has in homeland security applications, and in [14] how a similar system can work in variable lighting conditions.

In [15] Pulecio et al. discuss the use of image quality assessment for face images in the infrared spectrum. They recognize the impact of common image distortions on infrared face recognition and propose a method to improve identification rates by aggregating perceptual quality-aware features. Results of their study show that their method is robust against image distortions and applying image quality assessments prior to recognition experiments increased the overall performance of the matching system.

Likewise, Robila [16] discusses the use of infrared images in face detection, but goes on to discuss that pairing infrared imaging with the visible spectrum, to make up a hyperspectral range, is a more natural choice for face recognition because it is able to provide information beyond the normal visible spectrum, and would therefore exceed human sensing which could give computer vision an upper hand. Although their system was not as complex as typical commercial biometric applications, their experiments show that infrared images provide a valuable complementary database to use in face recognition.

2 Methodology

This section provides details of the data collection, facial recognition, and quality assessment efforts performed for this work. An indoor data collection was undertaken to capture visible-wavelength RGB facial images from 500 participants at a distance of 2 meters, using both unfiltered (ground truth) and filtered DSLR cameras. In addition to variations in filter wavelength, variable lighting was used to introduce inter-session variability. Matching experiments were performed using the NeuroTech VeriLook face matcher to evaluate the impacts of filters and lighting variation on template creation and match score values. Quality assessment of band filtered images to determine impact of band filtering on overall image quality were performed using machine learning techniques to perform face detection and quality assessment based on 5 different quality metrics: contrast, brightness, focus, sharpness, and illumination. The following sections provide more details for each of these aspects of the study.

2.1 Data Collection

The data collection (WVU IRB # 1605114472) was performed using protocols developed at WVU for the collection of SAP5/51 mugshot photos, with modifications made to meet the needs of this effort. A schematic view of the live subject-capture setup is shown in Fig. 1.

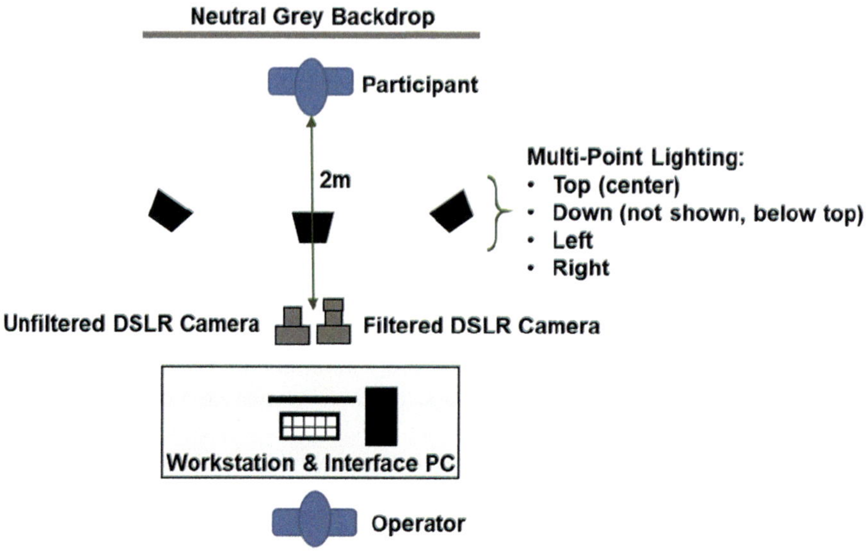

Fig. 1 Arrangement of image collection equipment used to collect images at 2 m

A Wescott neutral grey cloth backdrop was placed behind the participant who was seated on an adjustable-height armless seat. Standard 3-point tungsten lighting was used to provide uniform illumination on the face for ground truth images. A fourth light fixture ('Down' in Fig. 1) was included for variable lighting image capture, as described in the following Sect. 2.1.1. Two different camera/lens combinations were co-located on a camera stand situated 2 m away from the subject. These were as follows:

1. A Canon 5D Mark III DSLR camera, including: (a) a Canon Electro-focus (EF) 70-200 mm f/2.8 L, (b) Image Stabilization (IS) II and (c) an Ultrasonic Motor (USM) telephoto zoom lens for ground truth visible face images
2. A Canon 5DS R DSLR camera, including (a) a Canon Electro-focus (EF) 200 mm f/2 L, (b) Image Stabilization (IS) II, and (c) an Ultrasonic Motor (USM) telephoto lens used to capture band-filtered images

The camera/lens combination chosen in item (2) in the list above was based on the improved low-light performance of the Canon 5DS R camera body (better for narrowly-filtered imaged with lower light intensity) and the ability of the 200 mm lens to accept standard 52 mm diameter filters. A selection of bandpass filters was purchased for this camera/lens combination from Andover Corp. based on the range of visible bands reported in the literature to have better facial recognition performance. The filters actually used in the collection were selected based on the amount of light entering the 5DS R camera. Because wavelengths above 670 nm made the images too dark, the selected filters were as follows:

Filter	λ
0	Unfiltered
1	550±10
2	550±20
3	570±10
4	590±10
5	610±10
6	630±10
7	650±10
8	650±20
9	670±10

Fig. 2 Co-located arrangement of hyperspectral ground truth (GT) visible, and filtered visible imagers. The filter list provides an indication of the dominant colors of each wavelength

- 10 nm bandpass filters centered at 550 nm, 570 nm, 590 nm, 610 nm, 630 nm, 650 nm, and 670 nm
- 20 nm bandpass filters centered at 550 and 650 nm

Figure 2 shows the co-located cameras and the selected filters and their wavelengths.

The Canon DSLR cameras were connected to a PC via micro USB cables, and the hyperspectral camera was connected via a Camera Link cable and video interface card installed in the tower PC. The collection process from the multiple camera arrangement was controlled using a custom Collection Manager interface designed to allow operation of the Canon cameras (via an interface developed using the Canon SDK).

2.1.1 Collection Scenarios

Initially, facial images were captured with uniform 3-point lighting for angles varying from left to right profile in 22.5-degree angular yaw increments. In addition to these yaw angles, pitch angle was adjusted by having the participant look up and down. However, upon initial evaluation of these images it was discovered that

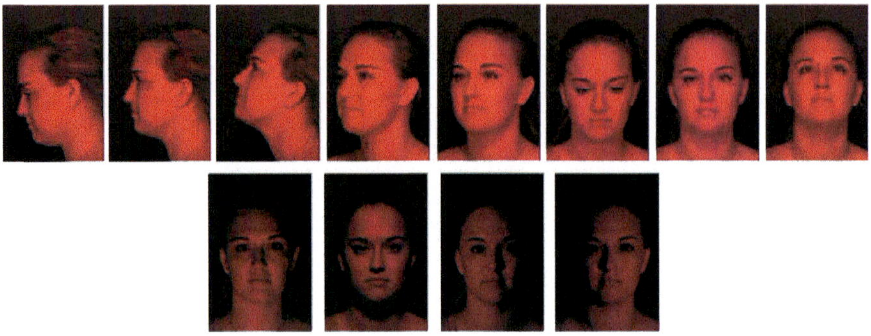

Fig. 3 Sample images for filtered face collection. *Top Row*: Angular variations with 3-point lighting. *Bottom Row*: frontal pose with single-point lighting

intersession variability was not significant enough. As such, the collection was changed to collect 17 different images from each participant with both angle and lighting variations, as follows:

- Three images at yaw rotations of −90, 90, and 0 degrees, with head tilted up, head tilted down, and no pitch rotation (9 images total)
- At the 0-degree yaw rotation position (frontal), 4 additional images were captured with lighting variation, using only on light to illuminate the face:
  - Top-light-only, left-light-only, right-light-only, bottom-light-only
- One image at 22.5, −22.5, 45, and −45 degrees yaw rotation with no pitch variation (4 images total).

Sample images from right profile to frontal pose at one filter wavelength are shown in Fig. 3.

2.1.2 Image Post-processing

In order to normalize the variations in pose angle in the images, a manual image cropping and co-registration tool was developed. The main functions of this tool are: (1) co-registering images based on pupil location, (2) normalizing pose to remove roll angle (if any). A view of the interface is shown in Fig. 4.

2.1.3 Data Collection Demographics

Data was collected from 500 individuals over 8 months of collection activity based on an average of 12 appointment slots per day. Table 1 provides a breakdown of the age, gender, and ethnicity of participants.

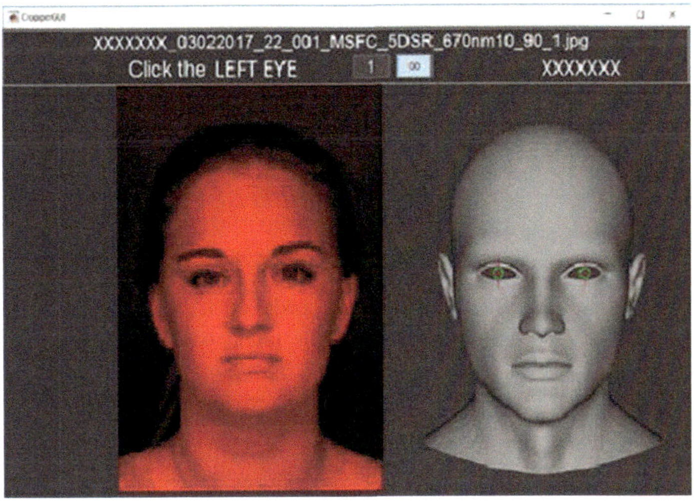

Fig. 4 Image cropping and co-registration interface

Table 1 Demographic information

Age	Ethnicity	Gender (# for each ethnicity)
18–19 – 18.4%	Caucasian – 66.4%	M – 161
		F – 171
20–29 – 68.6%	Asian Indian – 8.6%	M – 25
		F – 18
30–39 – 8.6%	East Asian – 8.2%	M – 22
		F – 19
40–49 – 1.6%	Middle Eastern – 4.4%	M – 12
		F – 10
50–59 – 1.4%	African American – 4.0%	M – 14
		F – 6
60–69 – 0.6%	Hispanic – 4.0%	M – 12
		F – 8
70–79 – 0.2%	African – 2.2%	M – 9
		F – 2
80–89 – 0.2%	Other – 1.6%	M – 3
		F – 5
Not reported – 0.4%	Pacific Islander – 0.6%	M – 2
		F – 1

The Caucasian demographic had the highest percentage of participation (66.4%), followed by East Asian/Indian totaling a combined 16.8%. Participation by other ethnicities ranged from 1.6% to 4%. The 20–29 age group had highest participation (68.6%), followed by the 18–19 group (18.4%) and the 30–39 group (8.6%). All other age ranges showed participation of 1.6% or less. The age and ethnicity distributions are consistent with the staff and student population of WVU, from which the majority of participants were recruited. Male participation outweighed male participation for all ethnicities except the 'Other' category.

2.2 Matching Evaluation of Band-Filtered Facial Images

Matching experiments were performed to determine any notable differences of genuine and impostor probability distributions produced from a commercial matcher between datasets of frontal pose face images taken using different visible light filters and under different lighting conditions. The commercial matcher used was the Neurotechnology MegaMatcher 6.0 (Revision #158433) with the VeriLook 6.0 SDK. Using the SDK, faces are detected followed by template creation using proprietary algorithms. Template information is internally processed and saved to '.dat' files, which are read by the Neurotechnology client for matching. The user can set a matching threshold to tailor the true matches based on their application. For this work, the threshold was set to zero so all match scores would be reported.

2.2.1 Matching Experiment Details

The image data used to create the matching gallery consisted of five front-facing images per subject. These images include a uniformly illuminated face (0_2) and one image each for lighting only on the left side of the face (LEFT), right side (RIGHT), above the face (TOP), and below the face (DOWN). Sample images are shown in Fig. 5.

The probe images consisted of filtered indoor face images grouped by the filter wavelength used on the camera when the image was taken. Each filtered dataset is identified by the wavelength and bandwidth size. For example, the dataset 550 nm10 represents the images taken with a 550 nm wavelength with 10 nm bandwidth filtered lens, whereas the 550 nm20 dataset contains images using the 550 nm wavelength with 20 nm bandwidth filtered lens. Following this scheme, we defined

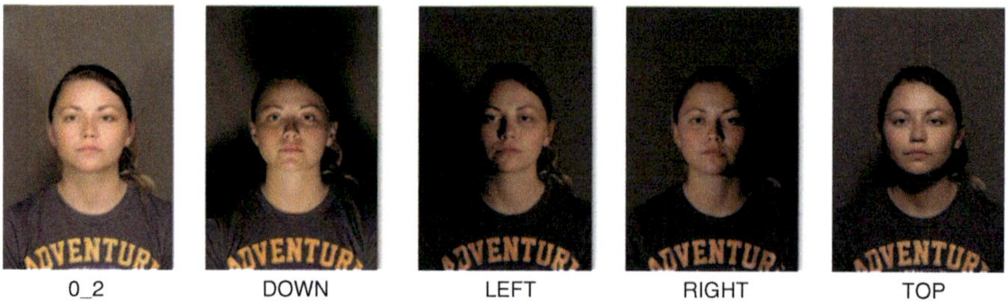

Fig. 5 Sample gallery images for one individual

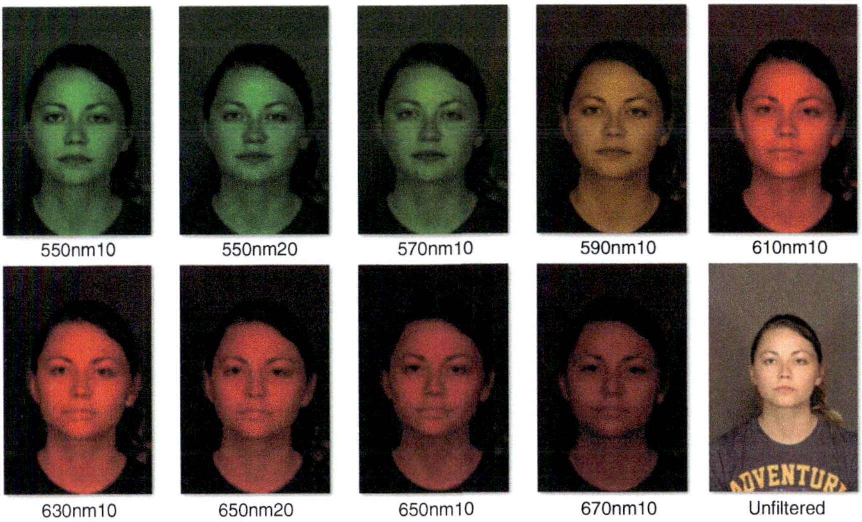

Fig. 6 Sample probe images for each wavelength for one individual

all the datasets as 550 nm10, 550 nm20, 570 nm10, 590 nm10, 610 nm10, 630 nm10, 650 nm10, 650 nm20, 670 nm10, and Unfiltered, which contains images taken with no filter. Sample images are shown in Fig. 6.

2.2.2 Matching Experiment Results

The impact of band filtered images used as probes in the VeriLook matcher can be observed via analysis of template creation. Figure 7 provides a summary of template creation and failure for each wavelength, with Fig. 8 providing further detail on template creation and failure for each lighting type. While processing templates using the Neurotechnology VeriLook 6.0 SDK, increasing the filter wavelength led to fewer templates created, with as few as 10% of the images passing template creation for the longest filter wavelength of 670 nm. Figure 8 highlights the differences in template creation for images with nonuniform lighting, with top, down, left, and right directional lighting causing template creation to fail significantly, especially for band-filtered images. In general, images filtered to 590 ± 10 nm were most likely to pass template creation for this matcher.

Figure 9 provides examples of images that passed and failed template creation for each filter wavelength.

The images in Fig. 9 illustrate that, supporting the result in Fig. 8, variations and non-uniformity in lighting impact template creation. However, it can also be seen that band-specific image contrast due to skin tone can also impact template creation for images captured under uniform illumination as well.

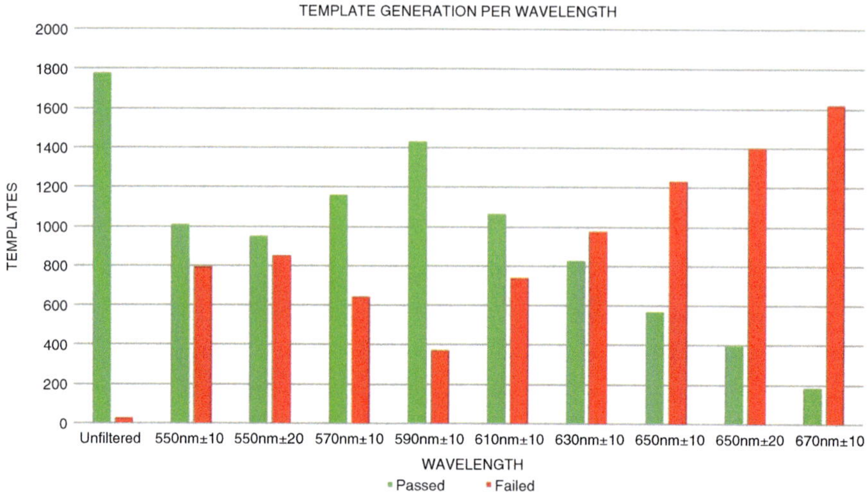

Fig. 7 Template generation success/failure for all images in each filter wavelength

Fig. 8 Template generation success/failure lighting variation in each filter wavelength

The CMC curves for images that passed template creation for all wavelengths and light source variations are shown in Figs. 10 and 11. In each case, probe images consisting of the band-filtered images captured at each wavelength under different forms of illumination were matched against a gallery of uniformly-lit unfiltered images. Matching results of unfiltered probes matched against the unfiltered gallery are provided in each plot as a baseline. These results indicate similar recognition performance for uniformly-illuminated (three-point) lighting and left- and right-

Fig. 9 Example images for each wavelength that both passed and failed template creation

illuminated face images. The results for bottom-illuminated facial images indicate that 550 nm, 590 nm, 610 nm, and unfiltered images showed poorer matching performance compared to other bands. The results for top-illuminated images indicate that 610 nm, 630 nm, and 650 nm showed poorer performance compared to other bands. It should be noted that the unsuccessful template creation for images with nonuniform illumination across all wavelengths (as described in Fig. 9) add bias to these results by greatly reducing the number of images that were able to be tested.

The results presented in this section indicate that applying passband filters to RGB images does not produce a significant performance improvement in facial recognition matching results. Combined band filtering and illumination variation causes a high percentage of template creation failure as the filter wavelength is increased, with as few as 10% of the images passing template creation for the longest filter wavelength of 670 nm. One interesting observation from this study was that band-filtered images captured at 590 ± 10 nm were most likely to pass template

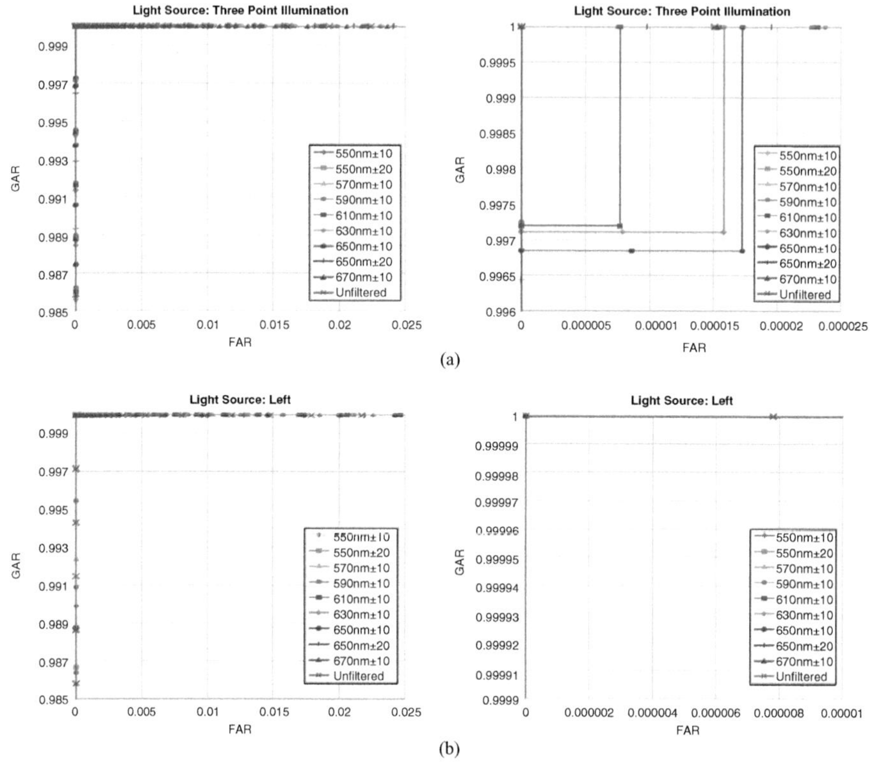

Fig. 10 CMC curves for images that passed template creation for (**a**) uniform 3-point illumination and (**b**) left-side illumination only for all filter passbands. The right column is a 'zoomed in' version of the image in the left column

creation for this matcher, even in cases of non-uniform illumination. This indicates that band filtering may have a higher impact on overall image quality rather than base match scores.

2.3 *Image Quality Assessment*

The performance of face recognition systems can be greatly impacted by the quality of the input image. Significant work has been completed to combat this issue for images in the typical visible band, but very little has been done for other wavelengths or filtered images. In this section we will discuss the Image Quality Assessment we have used on the data collected for this project and the parallels it has shown between image quality and face matcher performance for hyperspectral images.

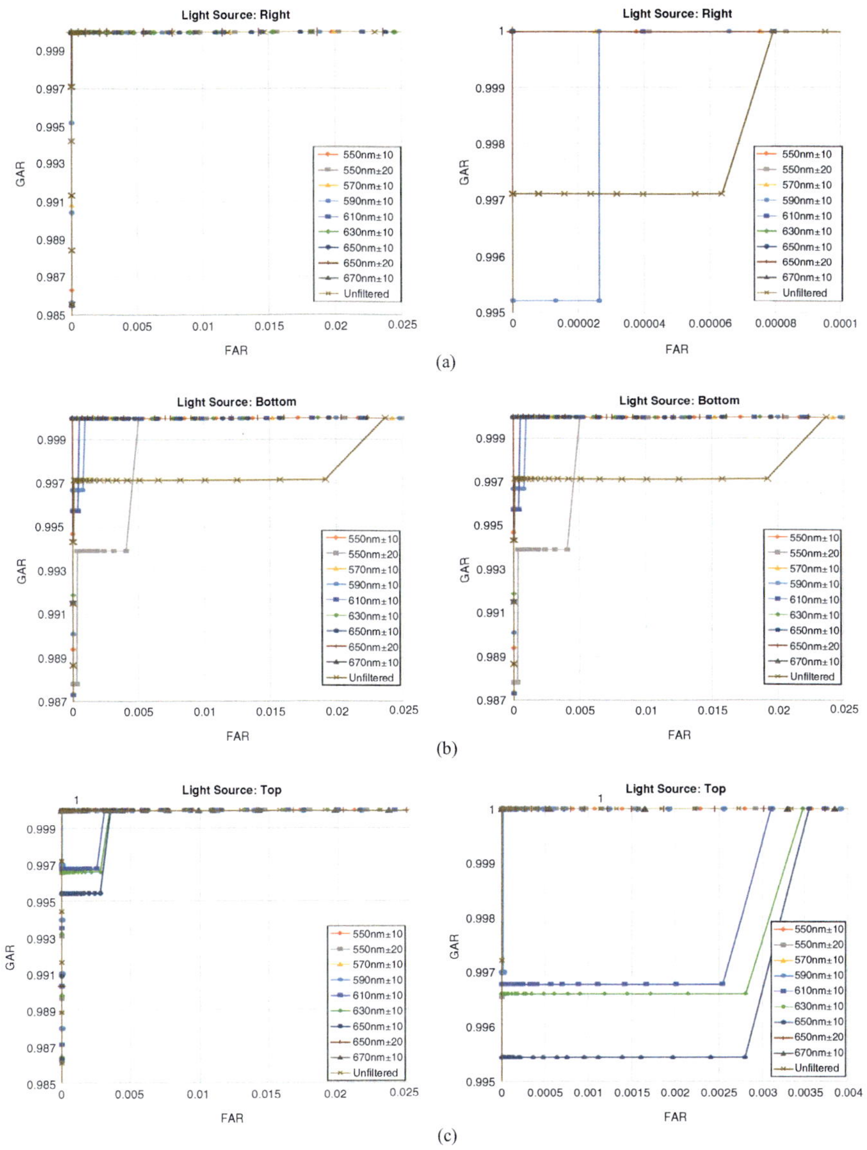

Fig. 11 CMC curves for images that passed template creation for (**a**) right-side illumination only, (**b**) bottom-up illumination only, and (**c**) top-down illumination only for all filter passbands. The right column is a 'zoomed in' version of the image in the left column. (Note: No scale change in (**b**))

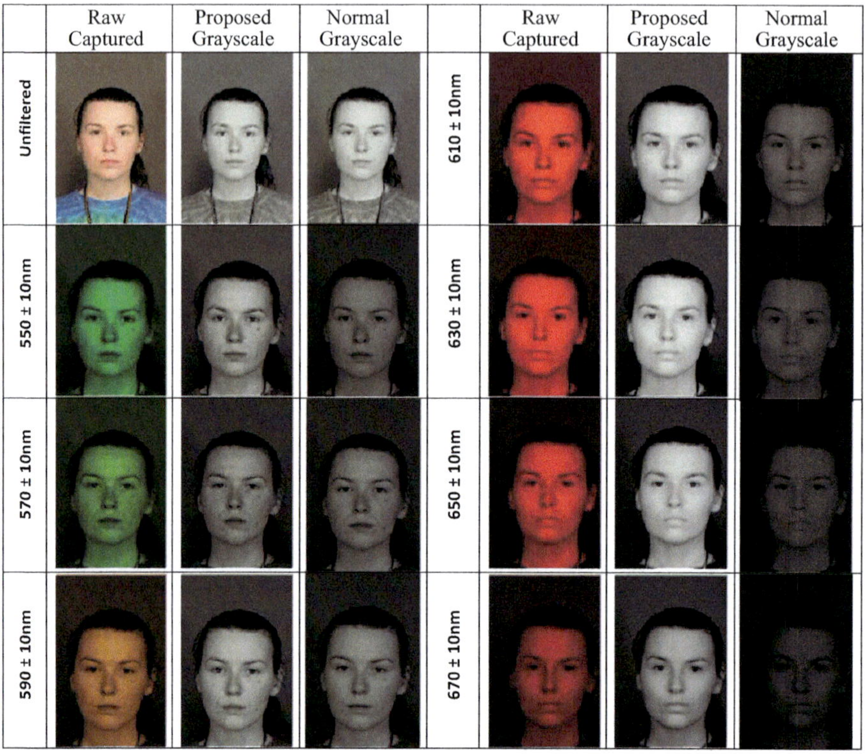

Fig. 12 Example images for each wavelength demonstrating the difference between the normal grayscale conversion method and our proposed method

2.3.1 Image Quality Assessment Experiments

The frontal-pose, uniform (3-point) illumination, indoor multispectral images captured using the Canon 5DS R camera were analyzed to determine if image quality is varied across the wavelengths collected.

- First, the images were cropped and converted to greyscale
- Then they were processed using a Multi-Task Cascaded Convolutional Neural Network (MTCNN) to perform face detection

Built-in grayscale conversion in MATLAB was not sufficient for this problem since there was not an equal distribution of Red, Green, and Blue pixels, so we developed our own method. The comparison of images can be seen in Fig. 12.

The first step was to separate the image into three separate images, one containing only red values, one containing only green values, and one containing only blue values. From here, the pixel intensities were found by determining the total number of Red pixels, Green pixels, and Blue pixels. The following formulas were then computed to find the true grayscale image.

Table 2 Face detection percentages for each imaging wavelength

Wavelength	Total number of faces	Number of detected faces	Percentage
550 nm	3654	3456	94.58
570 nm	3626	3472	95.75
590 nm	3626	3563	98.26
610 nm	3626	3592	99.06
630 nm	3626	3601	99.31
650 nm	3624	3581	98.91
670 nm	3613	3464	95.88
Norm	3624	3620	99.89

$$Total = Value_{Red} + Value_{Green} + Value_{Blue}$$

$$Avg_{color} = \frac{Value_{color}}{Total}$$

$$Grayscale{=}Avg_{Red} * Image_{Red}{+}Avg_{Green} * Image_{Green}{+}Ave_{Blue}{*}Image_{Blue}$$

The number of faces detected at each wavelength are presented in Table 2.

These results indicate that the 630 nm-filtered images exhibit the highest face detection performance, with all filter wavelengths performing above 94%.

After face detection was completed, the images were processed using a quality assessment tool.

Our image quality assessment tool computes values for contrast, brightness, focus, sharpness, and illumination based on Abaza et al. work in Design and Evaluation of Photometric Image Measures for Effective Face Recognition. We ran the image quality assessment over all frontal face images from the data collection mentioned above (29,019 images).

Contrast is defined as the difference in color intensities that makes the face distinguishable. We used the following equation [17] to find face contrast in our tool

$$C_{RMS} = \frac{\sum_{x=1}^{M} \sum_{y=1}^{N} \left[I\left(x, y\right) - \mu\right]^2}{MN}$$

Where μ is the mean intensity of the face image represented by I(x,y) of size N × M.

Brightness in our tool is determined by the equations [18].

$$\begin{bmatrix} r \\ g \\ b \end{bmatrix} = \frac{1}{255} \times \begin{bmatrix} R \\ G \\ B \end{bmatrix}$$

$$B = \frac{1}{MN} \sum_{x=1}^{M} \sum_{y=1}^{N} [\max(r, g, b)]$$

The first is the normalization of RGB (red, green, blue) to a range of 0–1 and the second is the conversion from RGB values into HSB (hue, saturation, and brightness). Brightness is an important attribute because it determines whether a visual stimulus appears to be more or less intense.

Focus refers to the degree of blurring of face images in our toolbox we compute two measures for focus that were originally proposed by Yap and Raveendran [19], the L_1-norm of the image

$$F_{L_1} = \sum_{x=1}^{M} \sum_{y=1}^{N} |G_{xx}(x, y)| + |G_{yy}(x, y)|$$

And the energy of the Laplacian of the image

$$F_{EL} = \sum_{x=1}^{M} \sum_{y=1}^{N} \left[G_{xx}(x, y) + G_{yy}(x, y)\right]^2$$

Where G_{xx} and G_{yy} are the second derivatives in the horizontal and vertical directions. We combine these measures and find the average of the two for our final focus measurement.

Image sharpness is defined as being the clarity of coarse and fine details in a face image. There are several sharpness measures that have previously been proposed, but for our tool we use an average of the following equations. The equation for S_1 was proposed by Kryszczuk and Dryhajilo [20] and S_2 was proposed by Gao et al. [17].

$$S_1 = \frac{1}{2}\left[\frac{1}{(N-1)M} \sum_{x=1}^{M} \sum_{y=1}^{N-1} \left|I_{x,y} - I_{x,y+1}\right| + \frac{1}{(M-1)N} \sum_{x=1}^{M-1} \sum_{y=1}^{N} \left|I_{x,y} - I_{x+1,y}\right|\right]$$

$$S_2 = \sum_{x=1}^{M-2} \sum_{y=1}^{N-2} G(x, y)$$

Where G(x,y) is the gradient value at (x,y).

The illumination measure [21] we chose to use is calculated using the following equation

$$I_2 = \sum_{i=1}^{4} \sum_{j=1}^{4} w_{ij} * \overline{I}_{ij}$$

Where an image is divided into (4 × 4) blocks and w_{ij} is the weight factor of each block.

Once values were computed for each category, normalization was completed. Initially we only used min-max normalization to see which wavelength performed best for each category, for this we used the following equation.

$$X_i = \frac{X_i - \min(X)}{\max(X) - \min(X)}$$

We found that overall 630 nm performed best across all categories. We then went one step further to create a fused score so that overall image quality could be assessed. In order to do this we first completed TanH normalization for all categories in order to normalize the distribution so that we can fuse the scores fairly. This is completed using the following equation

$$X_i = 0.5 * \left(\tanh\left(0.01 * Y_i - \frac{mean(Y)}{std(Y)} \right) + 1 \right)$$

We then use the mean fusion technique to fuse the scores into one.

$$Fused_i = \frac{Contrast_i + Brightness_i + Focus_i + Sharpness_i + Illumination_i}{5}$$

Once fusion is complete we use min-max normalization again to distribute the values between 0 and 1, with 0 being the worst quality image and 1 being the best quality image. This again showed that 630 nm produced higher quality images when compared to all other wavelengths.

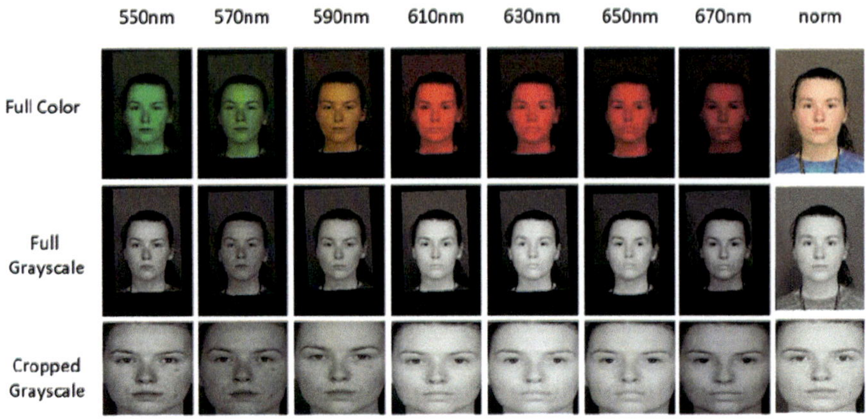

Wavelength	Contrast	Brightness	Focus	Sharpness	Illumination
550nm	0.895	0.839	0.625	0.830	0.869
570nm	0.861	0.593	0.442	0.627	0.592
590nm	0.858	0.893	0.547	0.680	0.830
610nm	0.888	0.885	0.532	0.631	0.952
630nm	0.921	0.924	0.475	0.570	0.960
650nm	0.916	0.926	0.503	0.624	0.945
670nm	0.899	0.864	0.642	0.776	0.851
Norm	0.786	0.903	0.515	0.474	0.899

Fig. 13 Top: sample face images before and after face detection. Bottom: Sample image quality results for a selected Caucasian female

2.3.2 Image Quality Assessment Results

Normalized results from each wavelength using min-max normalization are presented a single participant in Fig. 13, along with sample images of the individual.

In concordance with the face detection results, these images indicate that the 630 nm-filtered images possess higher overall quality scores for these two examples.

Average quality scores computed for all images in the indoor dataset for each quality category, along with a fused quality score that provides an overall image quality measure are provided in Table 3, and illustrated in a bar graph in Fig. 14.

These results show that the average and fused quality scores for all images are higher for the 630 nm-filtered face images. Box plots for each separate quality category, as well as the fused quality scores, are shown in Figs. 15 and 16.

The results indicate that contrast and sharpness generally have higher scores for all filtered images, with brightness, focus, and illumination having higher values for the 630 nm-filtered images.

Table 3 Average and fused image quality scores

Wavelength	Contrast	Brightness	Focus	Sharpness	Illumination	Fused score
550 nm	0.6470	0.2563	0.1580	0.3419	0.2695	0.4144
570 nm	0.6145	0.2029	0.1082	0.2840	0.2127	0.3576
590 nm	0.6948	0.2908	0.1694	0.3648	0.3092	0.4510
610 nm	0.7674	0.4309	0.3121	0.4918	0.4691	0.5955
630 nm	0.7748	0.4463	0.3348	0.5020	0.4909	0.6129
650 nm	0.7389	0.3658	0.2424	0.4219	0.4044	0.5282
670 nm	0.6125	0.2016	0.0941	0.2436	0.2246	0.3457
Norm	0.7730	0.4953	0.3604	0.5640	0.5219	0.6485

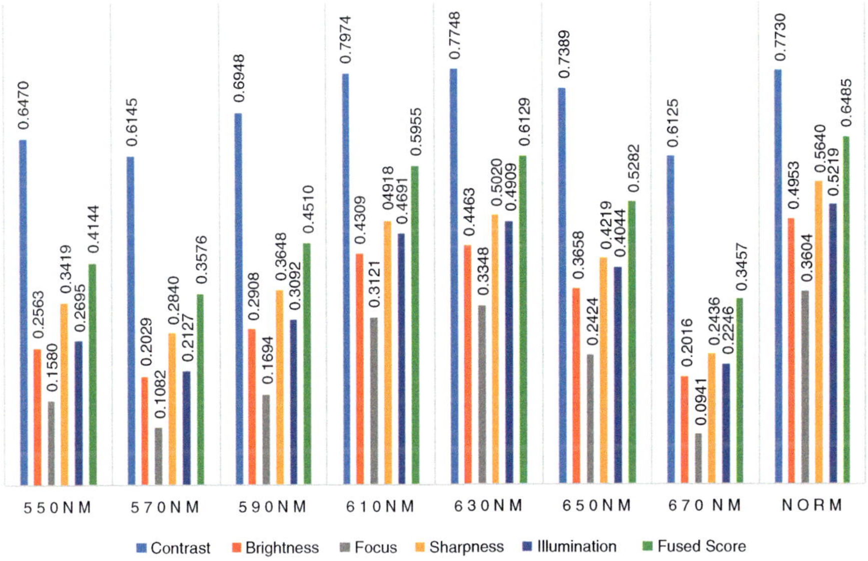

Fig. 14 Bar graph of quality scores presented in Table 2

3 Observations and Conclusions

Filtering RGB images does not produce the same effect as hyperspectral imaging systems, which do not employ color-specific pixels for imagery. The lack of significantly different matching performance of band-filtered RGB imagers is most likely due to the nature of the sensor. Because of the architecture of color CMOS imaging sensors, the filters simply provide a 'weighting' to either red, green, or blue pixels depending on the filter wavelength. This effectively decreases the amount of light reaching these pixels, and lowers overall image contrast. This phenomenon was clearly observed with longer wavelength filters (630 nm+) leading to a high degree of template creation failure for the matcher used in this study.

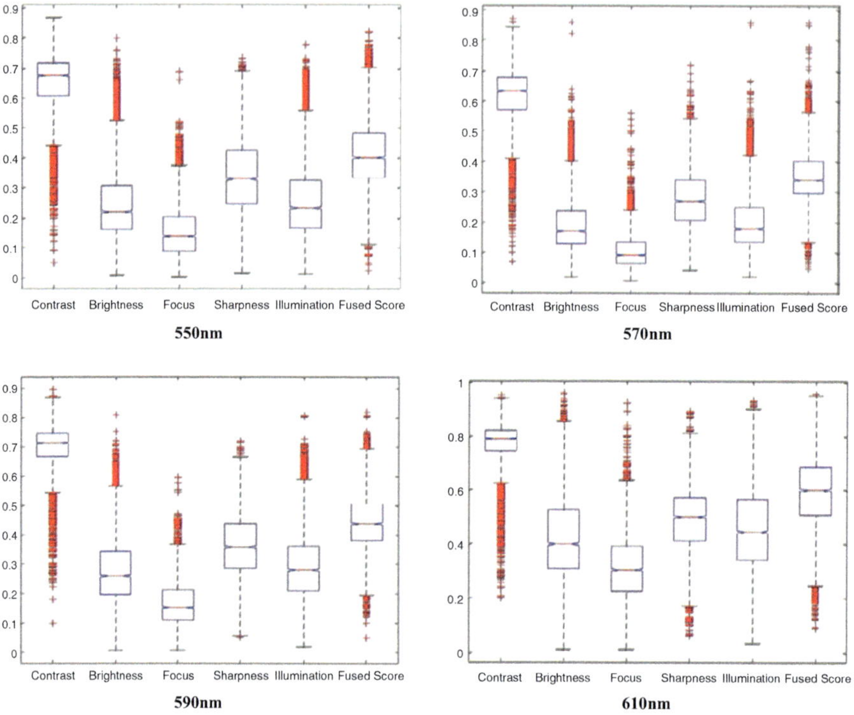

Fig. 15 Box plots of all quality categories for 550–610 nm filter wavelengths

Although these results are specific to the frontal-pose, uniform (3-point) illumination, indoor multispectral images captured using the Canon 5DS R camera, they indicate that band-filtered images possess varying quality within the range of wavelengths used in the data collection.

Contrast and sharpness quality categories seem to be the best quality metrics to use for band-filtered images, with images filtered to 630 nm possessing the highest quality scores, both in two specific examples extracted from the dataset as well as in the average quality scores across the entire dataset, both in each quality category and as fused scores.

While not an exhaustive exploration of operational scenarios, these results indicate that image quality may play a bigger role in the facial recognition performance of band-filtered images rather than simple band-filtering alone, warranting further study in this area.

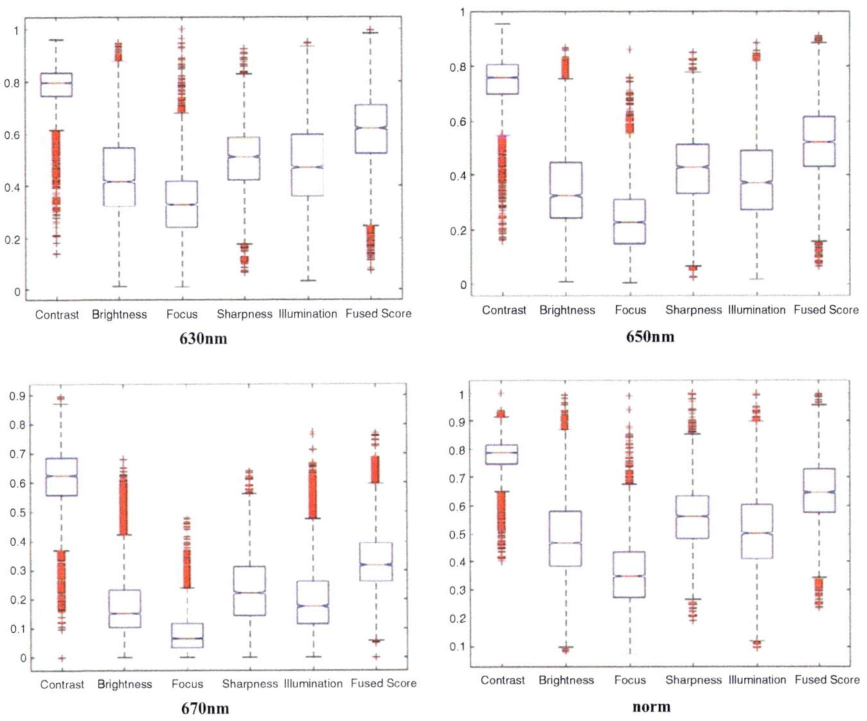

Fig. 16 Box plots of all quality categories for 630–650 nm filter wavelengths, as well as unfiltered (norm) images

References

1. Wang Z, Bovik AC, Lu L (2002) Why is image quality assessment so difficult? In: 2002 IEEE international conference on acoustics, speech, and signal processing, Orlando, FL, pp IV-3313–IV-3316
2. Wang Z, Bovik AC (2002) A universal image quality index. IEEE Signal Process Lett 9(3):81–84
3. Wang Z, Bovik AC, Sheikh HR, Simoncelli EP (2004) Image quality assessment: from error visibility to structural similarity. IEEE Trans Image Process 13(4):600–612
4. Sellahewa H, Jassim SA (2010) Image-quality-based adaptive face recognition. IEEE Trans Instrum Meas 59(4):805–813
5. Abaza A, Harrison MA, Bourlai T, Ross A (2014) Design and evaluation of photometric image quality measures for effective face recognition. IET Biom 3(4):314–324
6. Martin M, Bourlai T (2017) Enhanced tattoo image quality assessment through multispectral sensing. IEEE Sensors Lett 1(6):1–4, Art no. 7000404
7. Best-Rowden L, Jain AK (2017) Automatic face image quality prediction. arXiv preprint arXiv:1706.09887
8. Khryashchev V, Ganin A, Nenakhov I, Priorov A (2018) Improving audience analysis system using face image quality assessment. In: Favorskaya M, Jain L (eds) Computer vision in control systems-4, Intelligent systems reference library, vol 136. Springer, Cham

9. Khryashchev V, Nenakhov I, Lebedev A, Priorov A (2016) Evaluation of face image quality metrics in person identification problem. In: 2016 19th conference of Open Innovations Association (FRUCT), Jyvaskyla, pp 80–87
10. Chen J, Deng Y, Bai G, Su G (2015) Face image quality assessment based on learning to rank. IEEE Signal Process Lett 22(1):90–94
11. Di W, Zhang L, Zhang D, Pan Q (2010) Studies on hyperspectral face recognition in visible spectrum with feature band selection. IEEE Trans Syst, Man, Cybern Part A Syst Hum 40(6):1354–1361
12. Pan Z, Healey G, Prasad M, Tromberg B (2003) Face recognition in hyperspectral images. IEEE Trans Pattern Anal Mach Intell 25(12):1552–1560
13. Pan Z, Healey G, Prasad M, Tromberg B (2003) Hyperspectral face recognition for homeland security. In: Proceedings of SPIE 5074, infrared technology and applications XXIX
14. Pan Z, Healey G, Prasad M, Tromberg B (2004) Hyperspectral face recognition under variable outdoor illumination. In: Proceedings of SPIE 5425, algorithms and technologies for multispectral, hyperspectral, and ultraspectral imagery X
15. Pulecio CGR, Benítez-Restrepo HD, Bovik AC (2017) Image quality assessment to enhance infrared face recognition. In: 2017 IEEE International Conference on Image Processing (ICIP), Beijing, pp 805–809
16. Robila SA (2008) Toward hyperspectral face recognition. In: Proceedings of SPIE 6812, image processing: algorithms and systems VI, 68120X
17. Gao X, Li SZ, Liu R, Zhang P (2007) Standardization of face image sample quality. In: International Conference on Biometrics (ICB), Seoul, Korea
18. Bezryadin S, Bourov P, Ilinih D (2007) Brightness calculation in digital image processing. In: International symposium on technologies for digital fulfillment, Las Vegas, NV, USA
19. Yap P-T, Raveendran P (2004) Image focus measure based on Chebyshev moments. IEE Proc Vis Image Signal Process 151(2):128–136
20. Kryszczuk K, Drygajlo A (2006) On combining evidence for reliability estimation in face verification. In: European Signal Processing Conference (EUSIPCO), Florence, Italy
21. Abdel-Mottaleb M, Mahoor M (2007) Application notes – algorithms for assessing the quality of facial images. IEEE Comput Intell Mag 2:10–17

Unconstrained Face Recognition Using Cell Phone Devices: Faces in the Wild

Michael Martin and Thirimachos Bourlai

Abstract The ever growing field of face recognition is constantly expanding to tackle new and more challenging, problems as the advances in algorithms yield higher accuracy results. The most recent advances have opened up the possibility of conducting high accuracy face recognition on faces from completely uncontrolled sources, such as search engines, social-media, and other online sources. Conducting face recognition in this area is usually deemed as faces-in-the-wild, given the unbounded nature in which faces are collected. While performing face recognition on faces-in-the-wild datasets has many advantages, it can make it difficult to determine the limitations of the face recognition algorithm in terms of the scenarios in which the faces were collected. In this work, we will collect a simulated faces-in-the-wild dataset using four cell phones (common sources for faces-in-the-wild) in varying scenarios (distance, lighting, background, etc.) to fully demonstrate the capability of newly proposed deep learning based methods of face recognition. Furthermore, we will contrast this with previous, standard, methods of face recognition in the same scenarios to see how recent improvements in the filed have opened up new capabilities.

1 Introduction

Face recognition has many important security, military, and government applications that extend far beyond the base principle of being able to recognize a person by their face. While in many of these applications, an official or agent can control

M. Martin
West Virginia University, Morgantown, WV, USA
e-mail: mmarti40@mix.wvu.edu

T. Bourlai (✉)
Lane Department of Computer Science and Electrical Engineering, Multispectral Imagery Lab (MILab), West Virginia University, Morgantown, WV, USA
e-mail: Thirimachos.Bourlai@mail.wvu.edu

T. Bourlai et al. (eds.), *Securing Social Identity in Mobile Platforms*, Advanced Sciences and Technologies for Security Applications,
https://doi.org/10.1007/978-3-030-39489-9_7

the manner in which the face is collected, there exist some applications where manual control over face collection is not possible. Such applications may include surveillance, crowd monitoring, commercial store theft prevention and analytics, identification of combatants on a battlefield, with a near endless possibilities of use cases.

The process of face recognition has been heavily studied by the academic, industry, and government communities. Many methods have been proposed, with the successfulness being dependent on the use case and approach. Face recognition can generally be broken into two tasks of identification or verification [6]. In the instance of identification, a probe face image of unknown identity is matched to a set of gallery face images with known identity in the attempt to determine the identity of the probe image, also called 1-to-many matching. This is most commonly used in law enforcement applications when the identity of a individual is trying to be determined in the connection of a crime or illegal act. It is also used to check if an individual is on a watch-list, such as a no-fly list or a list of known terrorists. Face verification is used when the identity of a individual is combined with a probe face image and matched against the gallery entry for that individual for the purpose of confirming or rejecting that person's identity, also called 1-to-1 matching. It is most commonly used in security instances with cooperative subjects, but is not limited to this stipulation. We will perform experiments to explore both of these scenarios and demonstrate the accuracy differences.

Standard face recognition systems compare frontal facial images or probes, captured under controlled or challenging conditions with gallery, good quality face images to establish identity. These systems typically perform well when using good quality visible band cameras, when there is no illumination variation, and when subjects are cooperative and close to the camera. However, many law enforcement applications deal with mixed Face Recognition (FR) scenarios that involve matching probe face images, captured by different portable devices (cell phones, tablets, etc.), at variable distances and light conditions, against good quality face images (e.g., mug shots) acquired using high definition camera sensors (e.g., DSLR cameras). Although most portable devices operate in the visible band, the problem of cross-scenario matching, e.g., matching face images captured by different camera sensors and devices and at different conditions (indoor, outdoors, variable distances and illumination) is still an open area for research. This is also known as the heterogeneous FR problem and a potential solution will enable interoperability by adding a device-independent matching component.

Several works have focused on the use of faces-in-the-wild and the use of cell phone images for face recognition. One of the first works focusing on faces-in-the-wild proposed the use of a dataset captured from local new sources [7]. The use of datasets collected from online image search engines, such as Google and Bing, have also show to be of great advantage in constructed datasets of faces form uncontrolled sources [8, 20]. However, changes in the API for Google image search to limit the number of images that can be searched will likely hurt the creation of these types of databases in the future.

The exploration of mobile phones in biometrics has been of interest since the incorporation of cameras in their design [5]. The popularity of cell phones makes them a perfect medium from which a large number of images are collected by individuals. In 2016, there were an estimated 4.3 billion mobile phone users, covering 62.9% of people on the planet [15]. These numbers are only expected to increase, with a projected 67% by 2019, adding to the number of images being collected from mobile phones. For these reasons, the use of more advanced techniques in face recognition have begun to be explored in the academic community [9, 12, 18] and in industry as an additional security measure [11]. In this work, we will demonstrate the use of standard and state-of-the-art face recognition techniques on a simulated faces-in-the-wild cell phone dataset that was collected using common sensors in known, but varying, scenarios.

1.1 Cell Phone Image Capture

The image capture process in cell phones differs greatly from the image capture process in traditional cameras. The optics of cell phones are much less dependent on movement of lens and more dependent on software processing of the image. This is largely due to the form factor requirements of optics on mobile devices when compared to their traditional counterparts. In this work we describe traditional cameras as cameras that contain a distinct lens system, such as mirror-less and single-lens-reflex.

2 Cell Phone Face Database Creation

A database consisting of test subjects cell phone images in various scenarios was collected to test the performance of biometric algorithms. Data was collected for a total of 100 subjects, with a total of 80 videos being collected for each subject. A total of 4 cell phones were used, with 20 videos collected per phone. These 20 videos consisted of both indoor and outdoor conditions at various distances. In the instance of indoor conditions, both low lighting and full lighting scenarios were considered at a distance of 1, 5, and 10 m. Furthermore, image from both the front and rear cameras were captured for each scenario. For the outdoor, since the lighting could not be controlled, only one lighting scenario was considered. Demographic information for each test subject was also collected. This information can be seen in Fig. 1.

The four cell phones used in this collection were the Samsung Galaxy S4 Zoom, Nokia 1020, iPhone 5S and Samsung Galaxy S5. Videos were taken of each scenario with the subject rotating their head yaw from $+90^\circ$ to -90°. Test datasets were created from this database for the remaining test scenarios presented in this chapter.

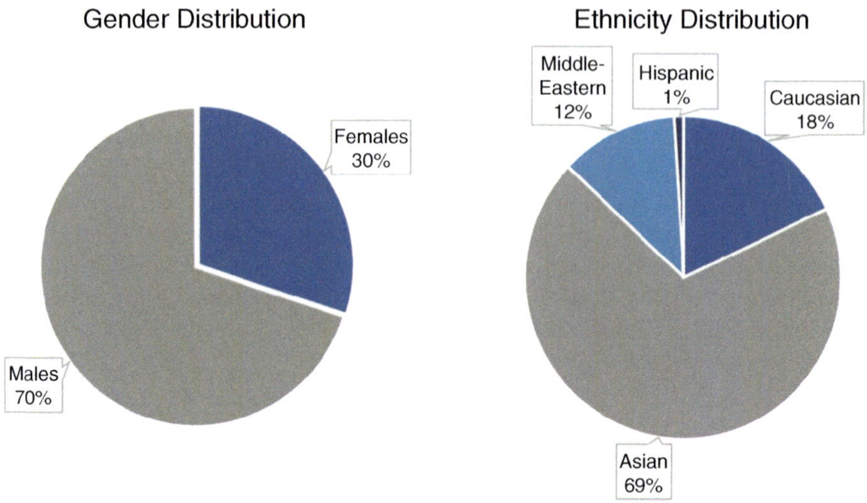

Fig. 1 Gender and ethnicity graphs of 100 subjects collected with cell phone cameras. The images were collected indoor and outdoor, at various lighting conditions, at distances of 1, 5 and 10 m, and with both the rear and front camera of the phone

3 Face Detection

The topic of face detection has been heavily explored in recent years leading to many breakthroughs in techniques and approaches. In this section we will demonstrate the performance of previously proposed traditional methods [17] and current state-of-the-art methods [1].

3.1 Traditional Methods

The most famous face detection algorithm of the past 20 years was proposed by Viola and Jones in [17]. This algorithm advanced the Adaboost learning method [3] used to train cascade based detection methods using more advanced features. In its traditional implementations, the algorithm places a bounding box on the objects it has been trained to detect, in this case, faces. Traditional implementations were not able to detect sub-facial landmarks (i.e. points on eyes, nose, mouth, etc.) without a separately trained cascade for each landmark, something that is possible with more modern techniques of face detection. Implementations of the Viola Jones algorithm (also often referred to as Haar-feature based cascade classifiers) are available with

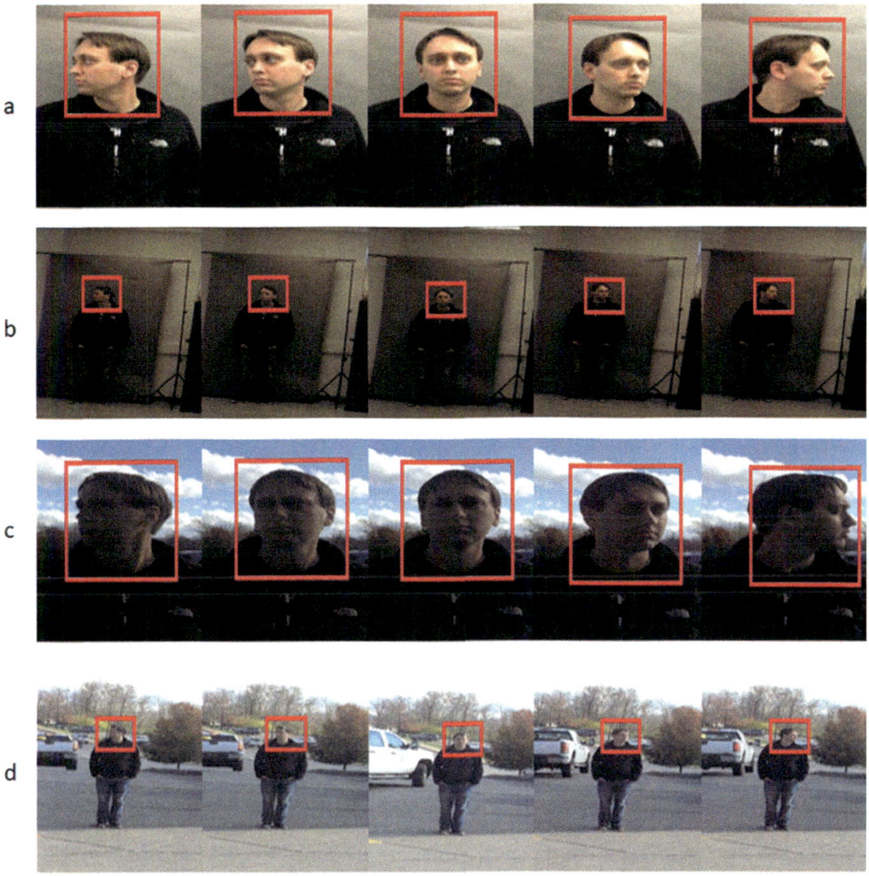

Fig. 2 Face detection demonstrated on images collected from mobile phones, captured under various conditions: (**a**) indoor-1 m, (**b**) indoor-10 m, (**c**) outdoor-1 m, (**d**) outdoor 10 m. The detected bounding boxes of the face are indicated in red

the popular Open Source Computer Vision Library (OpenCV[1]) for frontal face, profile face, and eye detection.

To demonstrate the capabilities of the Viola Jones algorithm, we tested its face detection on a test subset of cell phone images from our database. The testing subset consisted of 13 different scenarios for 100 subjects, 4 cell phones, distances of 1, 5, and 10 m, indoor and outdoor, and with a head yaw angle from +90° to −90°. The Viola Jones algorithm was found to perform well for indoor and outdoor conditions at close range, as shown in Fig. 2. However, long distances outdoors proved to be much more challenging and more errors were found. In face detection, there are two

[1]OpenCV: https://opencv.org/

Fig. 3 Successful cases of face detection via the Viola Jones algorithm on cell phone images are shown in "**a–d**", while "**e–g**" represent cases with false positives, in which a non-face region has been falsely identified as a face, and "**h**" demonstrates false negative, in which a face was falsely not identified

types of errors that can occur. These types of errors are False Positives, in which erroneous areas of the image are indicated as faces, and False Negatives, in which faces are not correctly indicated by the detection algorithm as faces. Examples of successful face detections and cases with errors are shown in Fig. 3.

Often, the Viola Jones is found to be insufficient to handle many of the challenging scenarios faced in today's atmosphere of faces-in-the-wild. Fortunately, many more advanced techniques have been developed that are capable of not only performing face detection more accurately in challenging scenarios, but also detect sub-facial features that can be used to locate areas of the face and pose estimation.

3.2 Face Detection with Pose Estimation

Many methods of face detection are capable of estimating the pose of the subject or detecting sub-facial features (i.e. eyes, noise, mouth, points on the jaw, etc.). This information is often important to filter out faces that are at extreme angles, or to extract information for certain regions of the face. To demonstrate this capability, we tested our cell phone dataset on a newly proposed method of face detection [19]. In this work the authors proposed a coupled technique combining cascade

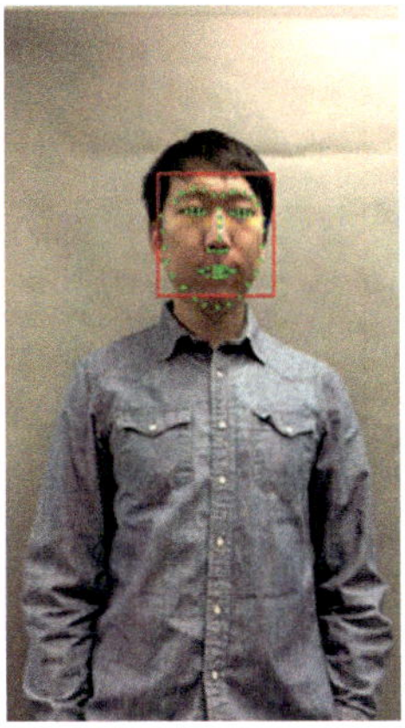

Fig. 4 Face Detection with Subfacial Features The detected face is indicated by a red box, with the subfacial features around the noise, jaw, eyes, and mouth indicated with a green plus

(Viola Jones) based detection with Convolutional Neural Networks (CNNs). An example detected face using this method is shown in Fig. 4. The sub-facial features are indicated by green points along the jaw, mouth, noise, eyes, and brow line.

Estimating the pose does not always require finding sub-facial features, many methods have been proposed that can classify face images into rough pose categories, such as profile (side of the face at a yaw angle of 90° or −90°) or frontal. There are three types of face pose rotation angles such as yaw, pitch and rolling angle. We used the algorithm developed by Aghajanian et al. [2]. to estimate the face pose, for the database collected under un-controlled conditions. This algorithm classifies the detected face images into three categories left profile, frontal and right profile, with yaw angle from −90°to 90°(see Fig. 5).

4 Face Recognition

In this section, we will describe several approaches to face recognition and demonstrate their use with images from our cell phone face dataset. Similarly to face detection, face recognition has many different approaches with many being variations on a fundamental approach. We will describe several different approaches with several different scenarios.

4.1 Previous Methods

In the early years of face recognition, the use of texture based algorithms were a common method of matching faces. Using texture based methods, hand crafted face features can be extracted that, when compared, allow the determination of a distance metric between two face images. These methods are used to get the

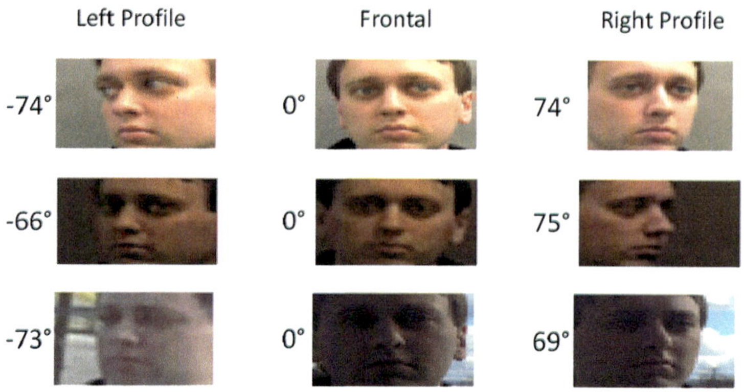

Fig. 5 Face Detection with Probe Estimation Face pose is estimated from −90° to +90° angles for the detected face images

appearance and texture information and is invariant to change in illumination conditions. Many variations of these algorithms exist, but the most popular texture algorithms proposed have been the Local Binary Patterns (LBP) [10] and Local Ternary Patterns (LTP) [21].

4.1.1 Simple Experiment

To test face recognition on our challenging-in-the-wild cell phone dataset with traditional methods, we used the LBP feature extraction algorithm in a series of face recognition experiments. Face videos were collected indoors, outdoors, at a standoff distance of 1 and 10 m. Data from 48 subjects in the collection was used to conduct the experiment. For each subject to run the experiments selected 4 samples per subject. To perform the experiments the probe and gallery sets were formed as described below:

- **Gallery Set:** Good quality neutral face images collected under controlled conditions were selected to form the gallery set. A single image was used per subject, resulting in a gallery size of 48 images.
- **Probe Set:** Face images from uncontrolled conditions, including indoors and outdoors were used to construct probe sets for each of the 48 individuals. A total of 16 samples from each phone (4 cell phones in total: 64 images in total for one subject) were used for each scenario of indoors/outdoors at a standoff distance of 1 and 10 m. This resulted in a total size of 3072 images for the probe set.

Texture based methods are particularly susceptible to inconsistencies in feature extraction due to the features being extracted from localized regions within the image. To help compensate for this, face images can be geometrically normalized, a process in which the images are rotated and scaled such that the face location

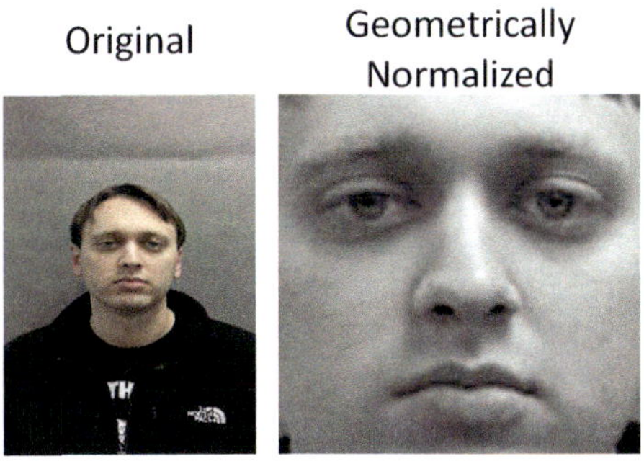

Fig. 6 Gallery Set Raw image (Left) and normalized image (Right)

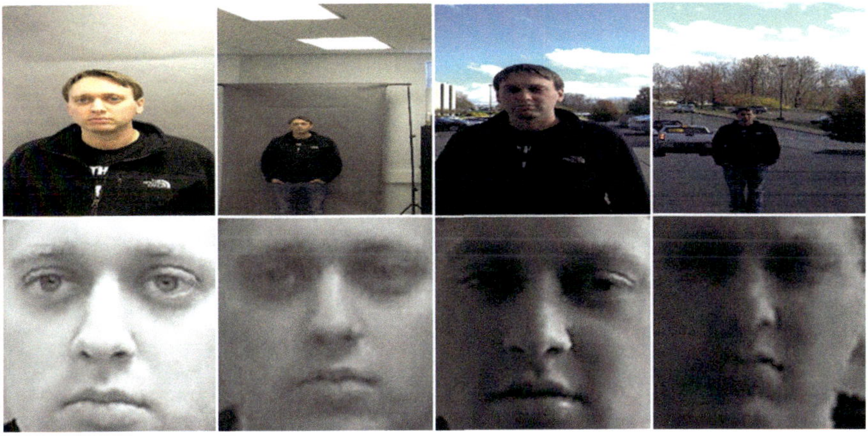

Fig. 7 Probe Set Face images in unconstrained light conditions including, Indoors 1 m, Indoors 10 m, Outdoor 1 m and Outdoors 10 m. Raw Images (Top) and Normalized Images (Bottom) with size of 111 × 121 Pixels

within the image is standardized. The geometric normalization scheme compensates for slight deviations in the frontal pose. It is composed of two main steps, eye detection and affine transformation. The eye center positions are first located by manual annotation and are used to geometrically normalize the images. Based on the located eye coordinates, the canonical faces were automatically constructed by applying an affine transformation. Faces are first aligned by placing the coordinates of the eyes in the same row such that the slope between the right and left eye is zero degrees. Finally, all the faces are canonicalized to the dimension of 111 × 121 pixels. The geometrically normalized images can be seen Figs. 6 and 7.

The results of our experiments are shown in Fig. 8 and Table 1. The results of the identification experiments are shown in the form of a Cumulative Match Characteristic (CMC) curve, where each the probably of identification in the number of top results is indicated on the horizontal axis as ranks. To elaborate, the probability of identification at a particular rank X corresponds to the probability that a correct match has a better or equal score to the Xth highest score. The results

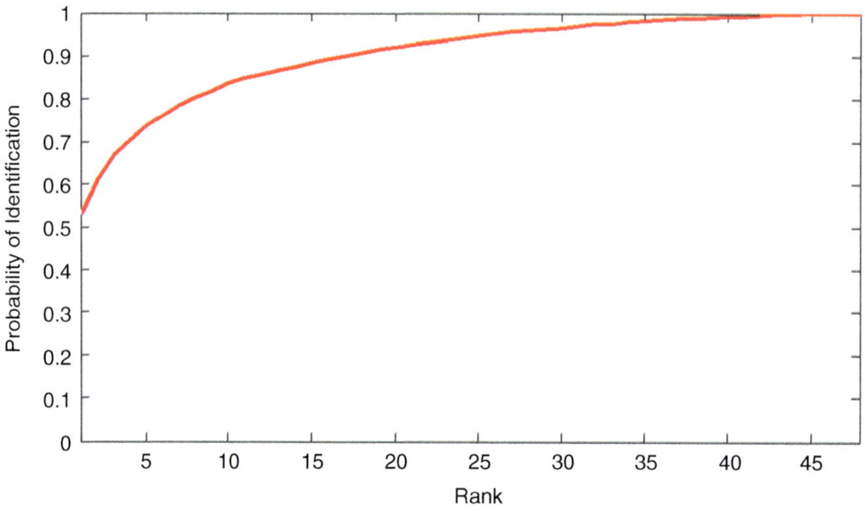

Fig. 8 Face Identification Results for faces-in-the-wild using basic matchers The Cumulative Match Characteristic (CMC) curve is shown for an identification experiment involving cell phone images with a gallery size of 48 images and a probe set size of 3072 in varying scenarios

Table 1 Results for LBP-CHI matcher for 1-m distance

	Rank 1	Rank 2	Rank 3	Rank 4	Rank 5
LBP-CHI	0.53	0.61	0.66	0.70	0.73

of our experiment indicate that traditional face recognition methods are not able to compensate for the challenges associated with faces-in-the-wild collected from cell phones and various conditions. The true identification rate (Rank-1 score) achieved was 53%, which is anticipated given the nature of the experiment.

4.1.2 Same Distance Matching

In the next experiment, we will slightly constrain the faces-in-the-wild nature of our dataset to images captured from the same distance and the same phone in order to observe the effect on our recognition rates. We will also use the TBP texture based feature extraction method in addition to LBP. The gallery and probe sets were created as described below:

- **Gallery Set:** The gallery consists of baseline images collected for each phone at the distances of 1 and 5 m. Faces images at 5 m were collected by manually zooming the digital focus of the camera on the face. For this experiment, two images for each of the 100 subjects, to make a total size of 200 images per phone and distance.

Table 2 Results for LBP-CHI matcher for 1-m distance

	Rank 1	Rank 2	Rank 3	Rank 4	Rank 5
iPhone 5S	0.95	0.96	0.96	0.97	0.97
Nokia 1020	0.82	0.87	0.89	0.89	0.90
Samsung Galaxy S5	0.89	0.92	0.95	0.96	0.96
Samsung Galaxy S4 Zoom	0.94	0.96	0.97	0.97	0.98

Table 3 Results for LBP-CHI matcher for 5-m distance

	Rank 1	Rank 2	Rank 3	Rank 4	Rank 5
iPhone 5S	0.70	0.73	0.75	0.77	0.77
Nokia 1020	0.63	0.69	0.72	0.74	0.75
Samsung Galaxy S5	0.68	0.73	0.77	0.78	0.80
Samsung Galaxy S4 Zoom	0.75	0.80	0.84	0.85	0.88

Table 4 Results for LTP-CHI matcher for 1-m distance

	Rank 1	Rank 2	Rank 3	Rank 4	Rank 5
iPhone 5S	0.96	0.96	0.96	0.97	0.97
Nokia 1020	0.81	0.87	0.88	0.89	0.90
Samsung Galaxy S5	0.90	0.92	0.95	0.96	0.96
Samsung Galaxy S4 Zoom	0.94	0.96	0.97	0.97	0.97

Table 5 Results for LTP-CHI matcher for 5-m distance

	Rank 1	Rank 2	Rank 3	Rank 4	Rank 5
iPhone 5S	0.70	0.74	0.75	0.76	0.87
Nokia 1020	0.64	0.70	0.72	0.74	0.76
Samsung Galaxy S5	0.68	0.73	0.78	0.79	0.80
Samsung Galaxy S4 Zoom	0.76	0.80	0.84	0.86	0.89

- **Probe Set:** Similarly, the probe set consists of 200 images with two images per subject for each of the 100 subjects at distances of 1 and 5 m.

The results of these experiments are shown in Tables 2, 3, 4 and 5 and are separated by distances.

4.2 *State-of-the-Art Methods*

While previous methods of face recognition may not be sufficient for faces-in-the-wild, recent advances in technology have opened up the possibilities of using uncontrolled face images in high accuracy scenarios. To fully demonstrate this we have repeated our experiment of same distance matching using newly proposed face recognition scheme, FaceNet [14]. This new method uses Convolutional Neural

Networks (CNNs) to extract deep facial features that exhibit much higher accuracy. The use of CNNs was first proposed in [4] and has become a very important algorithm in the modern community for image classification, feature extraction, and object detection. The base concept uses layers of image convolution with filters that have been optimized to extract ever more complex information as the number of layers grow. A more in depth understanding of the state of this technology can be found in [13]. In our instance, the output of the CNN would be a set of deep facial features that have been optimized for use with face recognition. These deep facial features are able to achieve high accuracy due to nature of being learned through annotated training data instead of hand-selected like previous techniques of feature extraction.

The first step in performing face recognition using these new methods is to crop our face images using a newly proposed algorithm, Multi-Task Convolutional Neural Network (MTCNN) [22]. This algorithm performs face detection much quicker and more accurately than the previously discussed methods, and normalizes the faces to be a square 160×160 image that can be used for face recognition with FaceNet. No geometric normalization is used as it is not essential to FaceNet that the images be geometrically normalized. In order to use FaceNet, we must first select the architecture we wish to use and train the model. Reported by the authors [14] as being the highest performing model, we chose to use the Inception-Resnet-V1 model proposed by Google in [16]. We trained the model with the CASIA face database [20] as described in the paper [14] to create the model which we will use for feature extraction in our recognition experiments. To perform the recognition experiments, cropped face images are fed through the network to produce a deep face feature vector. Feature vectors can then be compared using basic Euclidean distance to create distance metrics between two faces. Finally, we generate a distance matrix where an entry in the matrix $Dist(X_i, Y_j)$ corresponds to the distance score between the ith entry in the Probe set X and the jth Gallery entry in the set Y. This entire process completes the scheme for performing face recognition using this advanced deep learning methodology.

In order to test this method with the previously discussed methods, we devised a series of experiments for both face identification and verification. In the instance of identification, we repeated the same-distance matching experiment performed in Sect. 4.1.2 in which we matched 1 and 5 m images in varying scenarios. Using the distance matrix previously defined, we can easily compute the CMC curves and Ranks. The results of these experiments are shown in Tables 6 and 7. From these tables, we can see that the results are drastically different, with many of the scenarios achieving a 100% Rank-1 score for the true identification rate. The only instances where 100% was not achieved, was in the longer distance scenarios of 5 m with the iPhone 5S and Nokia 1020.

In the instance of verification, the distance matrix is still used but only the instances of the diagonal $Dist(X_i, Y_i)$ are considered, assuming that the indices of X_i and Y_i correspond to the same identity. The most common way of reporting accuracy for biometric verification, is the use of the Receiver Operator Characteristic (ROC) curve. This ROC curve shows the trade-off between the True Positive Rate

Table 6 Results for FaceNet for 1-m distance

	Rank 1 (%)
iPhone 5S	100
Nokia 1020	100
Samsung Galaxy S5	100
Samsung Galaxy S4 Zoom	100

Table 7 Results for FaceNet for 5-m distance

	Rank 1 (%)
iPhone 5S	99.0
Nokia 1020	95.0
Samsung Galaxy S5	100
Samsung Galaxy S4 Zoom	100

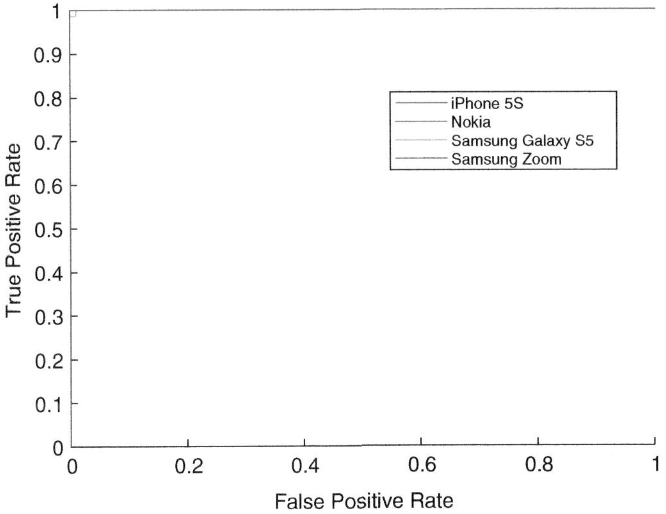

Fig. 9 Face verification Results for 1 m vs 1 m distance The Equal Error Rate (EER) for all the phones were found to be 0%, except the Nokia 1020 which had a EER of about 1%

(the rate at which a face is correctly verified as a matching identity) and the False Positive Rate (the rate at which a non-match is incorrectly verified as a matching identity). These results for the 1 and 5 m scenarios are shown in Figs. 9 and 10. The Equal Error Rates (EER), which correspond to when both the True Positive Rate (TPR) and False Positive Rate (FPR) are equal, if found to achieve a perfect result of 0% for all scenarios at 1 m except the Nokia 1020 (which achieved around 1% EER) The results were slightly worse at 5 m, where only the Samsung scenarios achieved a EER of 0%.

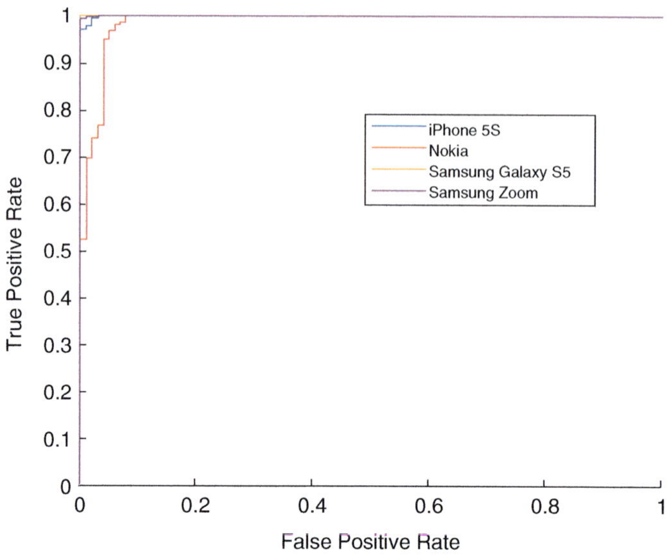

Fig. 10 Face verification Results for 5 m vs 5 m distance The Equal Error Rate (EER) for the Samsung Phones are found to be 0%, with a slightly higher EER for the Nokia 1020 and iPhone 5S

5 Summary

In this chapter we have discussed the uncontrolled nature of face recognition with faces-in-the-wild collected by cell phones and other devices. For that proposed we used a dataset of cell phone face images we collected in various controlled and uncontrolled scenarios. We have demonstrated the performance standard and state-of-the-art (available to us) face detection and face recognition methods. From the experiments, it can be seen that previous conventional FR techniques were not capable of achieving high accuracy (rank-1) in the uncontrolled scenarios seen with faces-in-the-wild. However, recent breakthroughs with deep learning have allowed the relaxing of these conditions such that we can now conduct face recognition in a variety of scenarios that previously were not possible.

Acknowledgements This research was supported in part by the Department of Homeland Security (DHS) and was conducted with the assistance of Dr. Neeru Narang.

References

1. Aghajanian J, Prince S (2009) Face pose estimation in uncontrolled environments. In: BMVC, vol 1, p 3
2. Aghajanian J, Prince SJD (2009) Face pose estimation in uncontrolled environments. In: In BMVC

3. Freund Y, Schapire RE, et al (1996) Experiments with a new boosting algorithm. In: ICML, vol 96, pp 148–156
4. Fukushima K (1980) Neocognitron: A self-organizing neural network model for a mechanism of pattern recognition unaffected by shift in position. Biol Cybernetics 36:193–202. https://doi.org/10.1007/BF00344251
5. Hadid A, Heikkilä JY, Silvén O, Pietikäinen M (2007) Face and eye detection for person authentication in mobile phones. In: ICDSC, pp 101–108
6. Jain AK, Li SZ (2011) Handbook of face recognition. Springer, Berlin
7. Learned-Miller E, Huang GB, RoyChowdhury A, Li H, Hua G (2016) Labeled faces in the wild: a survey. In: Advances in face detection and facial image analysis. Springer, Cham, pp 189–248
8. Liu Z, Luo P, Wang X, Tang X (2015) Deep learning face attributes in the wild. In: Proceedings of international conference on computer vision (ICCV)
9. Narang N, Martin M, Metaxas D, Bourlai T (2017) Learning deep features for hierarchical classification of mobile phone face datasets in heterogeneous environments. In: 2017 12th IEEE international conference on automatic face & gesture recognition (FG 2017). IEEE, pp 186–193
10. Ojala T, Pietikainen M, Maenpaa T (2002) Multiresolution gray-scale and rotation invariant texture classification with local binary patterns. IEEE Trans Pattern Anal Mach Intell 24(7):971–987
11. Robertson DJ, Kramer RS, Burton AM (2015) Face averages enhance user recognition for smartphone security. PloS One 10(3):e0119460
12. Rose J, Bourlai T (2019) Facial attribute estimation of mobile phone face data to aid in biometric systems. IEEE ASONAM (SNAST)
13. Schmidhuber J (2015) Deep learning in neural networks: an overview. Neural Netw 61:85–117
14. Schroff F, Kalenichenko D, Philbin J (2015) Facenet: a unified embedding for face recognition and clustering. In: Proceedings of the IEEE conference on computer vision and pattern recognition, pp 815–823
15. Statista (2016) Number of mobile phone users worldwide from 2015 to 2020. https://www.statista.com/statistics/274774/forecast-of-mobile-phone-users-worldwide/
16. Szegedy C, Ioffe S, Vanhoucke V, Alemi AA (2017) Inception-v4, Inception-Resnet and the impact of residual connections on learning. In: AAAI, vol 4, p 12
17. Viola P, Jones M (2001) Rapid object detection using a boosted cascade of simple features. In: Computer vision and pattern recognition, 2001 (CVPR 2001). IEEE computer society conference on proceedings of the 2001, vol 1. IEEE, p I
18. Wang L, Bourlai T, Dimitris M (2019) A coupled encoder- decoder network for joint face detection and landmark localization. IMAVIS 87(3):37–46
19. Wang L, Yu X, Bourlai T, Metaxas DN (2018) A coupled encoder-decoder network for joint face detection and landmark localization. Image Vis Comput 87:37–46
20. Yi D, Lei Z, Liao S, Li SZ (2014) Learning face representation from scratch. arXiv preprint arXiv:1411.7923
21. Yuan JH, Zhu HD, Gan Y, Shang L (2014) Enhanced local ternary pattern for texture classification. In: Huang DS, Bevilacqua V, Premaratne P (eds) Intelligent computing theory. Springer International Publishing, Cham, pp 443–448
22. Zhang K, Zhang Z, Li Z, Qiao Y (2016) Joint face detection and alignment using multitask cascaded convolutional networks. IEEE Signal Process Lett 23(10):1499–1503

Face Detection in MWIR Spectrum

Suha Reddy Mokalla and Thirimachos Bourlai

Abstract The capability to perform face recognition in the visible and thermal spectra is of prime interest in many law enforcement and military organizations. Face detection is an important pre-processing step for face recognition. Though many algorithms are available for face detection in the visible spectrum, an assessment of how these algorithms can be retrained for the thermal spectrum is an important study. Current available visible-based face detection algorithms are very effective in daytime conditions, however, when there are extreme changes in illumination conditions like very low-light to no light (night-time), these become challenging. Due to limited amount of data available for researchers from sensors in the thermal band (due to the increased cost of having and operating state of the art thermal sensors), there are only a few proposed algorithms. In this work, we conducted a study to determine the impact of factors such as indoor/outdoor environment, distance from the camera, application of sunscreen, training set size, etc. on training deep-learning models for a face detection system in the thermal spectrum that simultaneously performs face detection and frontal/non-frontal classification. Existing deep learning models such as SSD (Single Shot Multi-box Detector), R-FCN (Region Based Fully Convolutional Network) and R-CNN (Region Based Convolutional Neural Network), are re-trained using thermal images for face detection and pose estimation tasks. Results from each model are compared, and the model with the best performance is further trained and tested on different datasets, including indoor, outdoor at different stand-off distances. The highest accuracy is achieved using a Faster R-CNN model with ResNet-101 and the accuracy is 99.4% after a 10-fold cross-validation. More experiments are performed

S. R. Mokalla
West Virginia University, Morgantown, WV, USA
e-mail: sumokalla@mix.wvu.edu

T. Bourlai (✉)
Lane Department of Computer Science and Electrical Engineering, Multispectral Imagery Lab (MILab), West Virginia University, Morgantown, WV, USA
e-mail: thirimachos.bourlai@mail.wvu.edu; ThBourlai@mail.wvu.edu

T. Bourlai et al. (eds.), *Securing Social Identity in Mobile Platforms*, Advanced Sciences and Technologies for Security Applications,
https://doi.org/10.1007/978-3-030-39489-9_8

to further study the efficiency and limitations of this model. The data set we use was collected under constrained indoor and unconstrained outdoor conditions.

1 Introduction

Biometrics is the measurement and statistical analysis of people's unique physical and behavioral characteristics. The technology is mainly used for identification and access control, or for identifying individuals under surveillance. The basic premise of biometric authentication is that every individual can be accurately identified by his/her intrinsic physical and behavioral traits. There are several types of biometric modalities, including, but not limited to, fingerprint and retinal scanning, facial recognition and voice analysis. Face recognition holds high importance since it is non-intrusive, understandable, and can be collected in a covert manner at various stand-off distances. Multi-spectral face recognition, especially thermal-to-visible gained a lot of interest over the recent years due to the fact that thermal imagery best suits low light and nighttime conditions and face detection is an important pre-processing step for face recognition. Most of the face detection algorithms available in the open literature operate in the visible spectrum (390–700 nm). However, algorithms like eye detection, face detection, and pose estimation operating in the thermal spectrum are of much interest in many government and surveillance applications since the camera may need to be operated in a low-light to night-time zero visibility environment. The infrared (IR) spectrum plays an important role in such circumstances. The IR spectrum is divided into two categories – active IR and passive IR (also known as thermal). The active IR consists of Near IR (0.7–0.9 μm) and the lower range of Short-Wave IR (0.9–2.5 μm). During data acquisition in the active IR band, the subject's face is usually illuminated using an external light source. The passive IR consists of the Mid-Wave IR (MWIR) (3–5 μm) and the Long Wave IR (LWIR) (7–14 μm) bands. IR radiation in the form of heat is emitted from the subject's face and is detected by the camera sensor whenever data is acquired in the passive IR band. Passive IR sensors provide a significant capability of acquiring human biometric signatures under obscure environments without allowing the location of the sensor to be detected.

1.1 Related Work

There are many algorithms using deep learning and traditional approaches in the visible spectrum proposed in the open literature. However, the work done for face detection in the MWIR spectrum is limited. A couple of publications use local features combined with an AdaBoost classifier. Ma et al. [19] proposed a face detection algorithm that uses AdaBoost classifier with local features such as Haar-like, MB-LBP (Multi-Block Local Binary Pattern) and HOG (Histogram of

Oriented Gradients) and the task is accomplished by building stages by comparing the performance of different local features at each stage. Ma et al. [18] proposed two approaches based on local features – one is, they created new feature types by extending multi-block LBP and second is, an AdaBoost training method to get cascade classifiers with multiple types of local features, thus enhancing the description power of local features. Zheng [31] proposed face detection and eyeglass detection in thermal band using image denoising and normalization. Region growing algorithm is used for segmenting face in the image and a novel eyeglass detection method is proposed. Murata et al. [20] proposed a face detection algorithm that automatically extracts the target facial region from the thermal image by focusing on the temperature distribution of the facial thermal images as well as examine the automation of the evaluation. Reese et al. [22] presented and compared three face detection algorithms – Viola-Jones algorithm, Gabor feature extraction and classification using Support Vector Machines and a projection profile analysis using both Visible and LWIR spectra. They concluded that Gabor feature extraction can be re-trained using thermal images. However, the algorithm is extremely slow and projection profile analysis is applicable only to thermal images. Eveland et al. [6] proposed a human head tracking algorithm using three components – Method for modeling thermal emission from human skin that can be used for segmenting, segmentation model is applied to the condensation algorithm and tracking results are used to refine the segmentation estimate. Kwaśniewska et al. [15] re-trained the Inception *v3* model using thermal images (transfer learning) and utilized CNN localization ability to get information about classes' localization to detect and track faces in low-resolution thermal videos. Dowdall et al. [5] proposed a face detection algorithm in NIR spectrum in the following steps – frame acquisition, foreground-background segmentation, Near IR luminance calculation, Near IR illumination adjustment, skin detection and face detection.

There are a few algorithms that detect objects in the thermal spectrum other than face. Komatsu et al. [14] proposed a 3D imaging technique based on integral imaging. Using an LWIR camera multiple 2D images are captured that are known as elemental images of a scene with each image having a unique perspective of the 3D objects. A 3D scene is re-constructed and object detection using correlation filters and Support Vector Machines is performed. Herrmann et al. [2] proposed a person detection algorithm in which IR images are transformed as close as possible to visual RGB images and the remaining domain gap is corrected by training the CNN using a limited set of IR images. Biswas et al. [1] proposed a mid-level attribute in the form of multi-dimensional template using Local Steering Kernel (LSK) as low-level descriptors for detecting pedestrians in far IR images. In order to learn the LSK a new similarity kernel is introduced.

There are many face detection algorithms in visible spectrum available in the open literature using CNNs. Zhu et al. [32] proposed a Contextual Multi-Scale Based CNN with two contributions – the multi-scale information is grouped both in region proposal and RoI (Region of Interest) to deal with tiny face regions, second, the proposed network allows explicit body contextual reasoning inspired from the intuition of human vision system. Yang et al. [28] proposed Faceness-net, a new method to finding faces through scoring facial parts responses by their spatial structure and arrangement and the scoring mechanism is data-driven and carefully formulated to detect faces under severe occlusion and unconstrained pose variations. Jiang et al. [13] demonstrated state-of-the-art face detection results using the Faster R-CNN on three popular face detection benchmarks and compared different generations of region-based CNNs and a variety of other recent high-performing detectors. Ranjan et al. [21] proposed HyperFace, a deep multi-task learning framework that is used for simultaneous face detection, landmark localization, pose estimation and gender recognition and proposed two variants of HyperFace – 1. HyperFace-ResNet and 2. Fast-HyperFace. Sun et al. [24] proposed a face detection algorithm to improve the state-of-the-art faster R-CNN framework by combining a number of strategies, including feature concatenation, hard negative mining, multi-scale training, model pre-training and proper calibration of key parameters.

Zhang et al. [30] proposed joint face detection and alignment using Multi-Task Cascaded convolutional Neural Networks (MT-CNN) that exploits the inherent correlation between detection and alignment to boost up their performance. They leveraged a cascaded architecture with three stages and introduced an online hard-sample mining. Yang et al. [29] proposed a face detection algorithm using scale-friendly deep convolutional neural networks that could handle faces at extremely different scales. Farfade et al. [7] proposed a multi-view face detection algorithm, which they called Deep Dense Face Detector (DDFD), a method that does not require pose/landmark annotation and is able to detect faces in a wide range of orientations using a single model based on deep convolutional neural networks.

The goals of this work are *(1) to conduct an experimental study to determine the impact of training deep learning object detection models to detect faces in MWIR images and classify the faces into frontal and non-frontal (profile) simultaneously (2) to determine the most efficient model and suitable learning rates and (3) to determine in which conditions (training set size, scenario etc.) the performance starts degrading*. The existing deep learning models available for different tasks such as object detection, image recognition, face matching etc. are trained using millions of images. Training a CNN with many weights requires millions of samples, as well as High Performance Computing Sources and requires few days, sometimes, weeks for training. Transfer learning stands an acceptable alternative for this problem, where the model available for the required task is re-trained using a few hundred or thousand samples. It can be defined as re-utilizing the knowledge learned from one problem to another related one.

The goals listed are achieved by re-training and optimizing different deep learning models [12]. The dataset includes images collected at different distances (5 and 10 m) collected in both indoor and outdoor environments. The dataset also contains images in which sunscreen is applied to human subjects' faces. Application of sunscreen affects the vein pattern of the face, by penetration of nano particles of sun-screen into the epidermis [3]. Therefore, this may affect the information captured by the thermal camera. In this work, we propose a unified system that integrates face detection and classification using Convolutional Neural Networks (CNNs) into one nighttime face detection system. The work flow of the proposed system is shown in Fig. 1. First, MWIR frontal images are collected at distances of

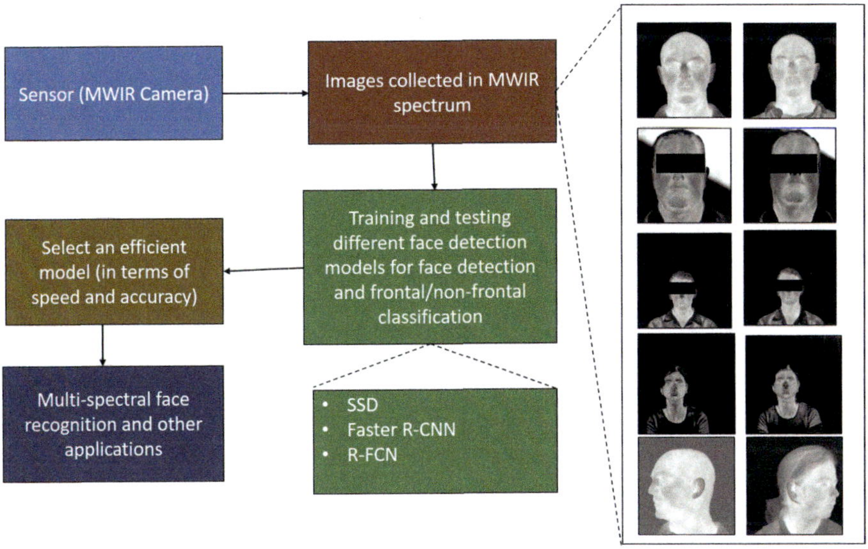

Fig. 1 Proposed system design for face detection and frontal/Non-frontal classification (Dataset Explained in 3 Section)

5 and 10 m, indoor and outdoor and non-frontal images are collected at 5 m, indoor. These images are used to train and validate the deep learning models by tuning the network (transfer learning) and the outputs from each model are compared. The best performing model is used to perform further face detection experiments described in the 2 section.

2 Methodological Approach

This section provides the theoretical framework of the fundamental restorative building blocks used that include conducting an empirical study to—(1) evaluate different models, and (2) determine the model that yields highest precision for the face detection and frontal/non-frontal classification. Understanding these key building blocks provides the reader an analytical foundation of each independent strategy. Additionally, this understanding also helps compliment the experimental testing and observations of each of the key building blocks and their combination described in the following section.

2.1 Deep Learning Models and Feature extractors

The deep learning models trained, tested, and optimized for face detection and frontal/non-frontal classification in this work are SSD, R-FCN, Faster R-CNN with MobileNet, Inception *v2*, Inception, *v3*, ResNet 101 as primary feature extractors.

All the aforementioned models are trained and tested using the MWIR data resulting in a total of approximately 100 experiments and the results summarized in 3 section. Meta architectures of the models are described below

- **SSD:** SSD [17] is a feed-forward CNN that produces a set of bounding boxes with confidence scores for the presence of object in those boxes. This is followed by a non-maximum suppression step that detects the object. The early networks are standard CNN layers that are used for classification in high quality images called base network and are truncated before any classification layers. Convolutional feature layers are added to the truncated base network that decrease in size and generates detections at multiple scales. Each added feature layer produces a fixed set of predictions using a set of convolutional networks which are different for each feature layer.
- **Faster R-CNN:** Faster R-CNN [23] is a relatively fast version of Fast R-CNN. It has two components. First is an RPN (Region Proposal Network) that takes image as an input and outputs a set of rectangular object proposals, each with score of the presence of object class in the box. To generate proposals, a small network is slid over the convolutional feature map output by the last shared convolutional layer and each sliding window is mapped into a lower-dimensional vector which is fed into two sibling fully-connected layers—a box regression layer and a box classification layer. The second component is the Fast R-CNN [8].
- **R-FCN:** The R-FCN [4] follows R-CNN in adopting the two-stage object detection strategy explained above. Candidate regions are extracted by RPN, after which R-FCN classifies the ROIs (Regions of Interest) into categories and background. The last layer of R-FCN is a position-sensitive ROI pooling layer that aggregates the outputs of the last convolutional layer. The backbone architecture of R-FCN is ResNet-101 which is explained later in this section. The significant change made in the ResNet-101 architecture is reducing the effective stride from 32 pixels to 16 pixels, increasing the score map resolution.

  Feature extractors are explained below
- **Mobilenets:** MobileNets [11] is basically designed for Mobile Vision applications and is a relatively faster convolutional network compared to the other models explained here. It is based on depth-wise separable convolutions which is a form of factorized convolutions which factorize a standard convolution into a depth-wise convolution and a 1×1 convolution called a point-wise convolution. The depth-wise convolution applies a single filter to each input channel and the point-wise convolution applies a 1×1 convolution to combine the outputs of the former.
- **Inception:** Inception [25], known popularly as GoogleNet, is the base architecture for Inception *v2* and *v3*, follows the basic idea to operate filters with multiple sizes on the same level. This model has 9 inception modules stacked linearly in 22 layers and uses global average pooling at the end of the last inception module. To address the problem of vanishing gradient [10], which is common in any very deep classifier, two auxiliary layers are introduced.

- **Inception *v2*:** Inception *v2* [26] is similar to GoogleNet with the following changes. Representational bottleneck is reduced as reducing the dimensions drastically may cause loss of information. To achieve this, filter banks in the module are expanded. 5×5 convolutions are factorized to two 3×3 convolutions and any $n \times n$ convolutions are factorized to $1 \times n$ and $n \times 1$ convolutions. This reduces the computational cost, as larger convolutions are extremely expensive than smaller ones.
- **Inception *v3*:** Inception *v3* [26] includes all the upgrades mentioned above for Inception *v2*. Additionally, the following details are added to the architecture — RMSProp optimizer, factorized 7×7 convolutions, Batch Normalization is applied to the auxiliary classifiers and label smoothing to prevent over-fitting.
- **ResNet 101:** ResNet [9] uses of referenced mapping instead of unreferenced i.e., the input from one layer is directly connected to the next layer along with the output from the previous layer. The intuition behind this approach is that it is easy to use a referenced mapping than it is to optimize a non-referenced mapping.

These networks are trained and tested using the thermal data explained in the above sections. The parameter tuning and corresponding results obtained are presented in 3 section.

2.2 *Further Experiments Using Faster R-CNN with ResNet-101*

Among all the models used, Faster R-CNN with ResNet-101 as the primary feature extractor yielded the highest accuracy. More face detection experiments are conducted using this model to further study the efficiency and limitations of the network. First group of experiments include changing the number of training images used i.e., the training set size is reduced by 10% each time. These experiments are performed to validate the network trained with a smaller number of images each time. The second set includes using data from one of the eight scenarios for training and data from the other scenarios to validate the data one after the other, for instance, data from 5 m, indoor without sunscreen is used for training and data from rest of the scenarios is used for testing. All the experiments performed are described in detailed in the next section.

3 Experimental Evaluation

An experimental study is conducted to assess the performance of the deep learning models for face detection in the MWIR spectrum. We froze all the parameters for the networks as described in their original works except the learning rate which is varied to tune the network and improve results.

Table 1 Number of images per category

Category	5IN	5IS	5ON	5OS	10IN	10IS	10ON	10OS
Number of images	795	780	795	690	810	795	765	777

3.1 Dataset Description

Number of subjects participated in the study is 56. The number of images in each of the eight categories is presented in Table 1 and the scenarios are explained here.

1. 5IN – 5 m, Indoor, no sunscreen applied to subject's face
2. 5IS – 5 m, Indoor, sunscreen applied to subject's face
3. 5ON – 5 m, Outdoor, no sunscreen applied to subject's face
4. 5OS – 5 m, Outdoor, sunscreen applied to subject's face
5. 10IN – 10 m, Indoor, no sunscreen applied to subject's face
6. 10IS – 10 m, Indoor, sunscreen applied to subject's face
7. 10ON – 10 m, Outdoor, no sunscreen applied to subject's face
8. 10OS – 10 m, Outdoor, sunscreen applied to subject's face.

The non-frontal data is comprised of 56 subjects, each with approximately 12 images per pose, totaling 1400 images. All non-frontal data was collected indoors at a distance of 5 m. No pre-processing or augmentation techniques are used on this data.[1]

3.2 Initial Experiments to Compare Different Models

All of the data (from eight scenarios and non-frontal images) is used to train and test the networks. First, each image is manually annotated to extract a face bounding box for training and validation purposes. Next, training and validation is performed as described in the 2 section. We use 90% of the data from each scenario for training and 10% for validation. The resulting model detects faces and classifies them as frontal or non-frontal classes. The precision values for our models, the time it takes to detect one face, and the learning rate used are summarized in Table 2.

It can be observed from Table 2, that the SSD model with MobileNet resulted in an accuracy of 92% with a learning rate of 3e-6 and it is relatively faster than the other networks. It can also be observed from the table that SSD with MobileNet took 1 ms for single detection with a learning rate of 0.00003, whereas, SSD with Inception *v2*, R-FCN with Resnet 101 and Faster R-CNN with Resnet 101 took 12 ms for single detection. Comparing the accuracy for all the four networks used,

[1]Release of MWIR Face Dataset: This is currently a private database with availability determined on case-by-case basis. If interested in working with this database, please contact the corresponding author.

Table 2 Face detection accuracy of different models

Model	LR	Training time (hours)	Detection time (ms)	Precision (%)
SSD with MobileNets	0.004	did not converge	–	–
SSD with MobileNets	1e-4	1.5	1	62%
SSD with MobileNets	3e-5	1.75	1	74%
SSD with MobileNets	3e-6	2.15	1	92%
SSD with Inception *v2*	1.85	12	1.85	71%
SSD with Inception *v2*	2.5	12	2.5	86%
R-FCN with Resnet 101	1e-4	2.25	45	72%
R-FCN with Resnet 101	3e-5	3.5	45	91%
Faster R-CNN with Resnet 101	2	1e-4	60	77%
Faster R-CNN with Resnet 101	**3e-5**	**2. 5**	**60**	**100%**

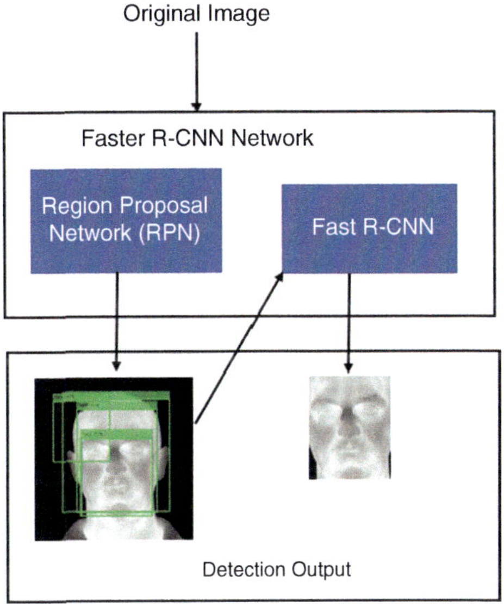

Fig. 2 Detection results using Faster R-CNN algorithm for an image at 5 m, indoor

it can be noticed that Faster R-CNN network with ResNet 101 network as feature extractor performed the best for our data with a learning rate of 0.00003.

3.3 Different Experiments Using Faster R-CNN Model with ResNet 101

Our initial experiments found that Faster R-CNN and 10-fold cross-validation yielded a maximum accuracy of 100% and average of 99.4% across all folds (Fig. 2). This model is further trained using different learning rates for different

Table 3 Accuracy of Faster R-CNN for different divisions of training and validation sets

(Training, validation)	Learning rate	Precision (%)
(80,20)	0.00003	100
(70,30)	0.00003	99.26
(60,40)	0.0004	98.72
(50,50)	0.0004	98
(10,90)	0.001	96.40

Table 4 Accuracy of Faster R-CNN for different indoor scenarios as training and validation sets (Rows: Validation Set, Columns: Training Set)

(Training, validation)	5IN	5IS	5ON	5OS	10IN	10IS	10ON	10OS
5IN	–	99.87	100.00	99.00	63.36	56.98	59.95	57.56
5IS	98.32	–	98.46	100.00	71.80	76.98	51.44	67.05
5ON	99.87	97.20	–	99.67	86.32	87.13	61.44	58.21
5OS	97.56	97.92	96.44	–	82.79	90.77	61.63	59.21
10IN	88.76	85.16	91.31	82.17	–	99.64	100.00	99.27
10IS	82.78	84.74	76.10	85.67	99.67	–	99.72	100.00
10ON	78.66	72.81	79.47	82.49	100.00	99.76	–	100.00
10OS	76.90	77.87	71.43	89.87	96.57	100.00	98.90	–

combinations of training and validation sets. The training data is decreased by 10% for each experiment until only 10% data is available for training. Results for these experiments are presented in Table 3.

It can be observed from Table 3 that the model performed well for all scales of data. The only change made from the previous experiments is the learning rate value. The learning rate is changed based on whether the model converged, the amount of time taken for convergence and accuracy observed on the validation set. Different learning rates are used for different sets of data and the results summarized in Table 3 are the best considering all the aforementioned conditions.

This model is further used to perform only face detection (not including the frontal/non-frontal classification) experiments using data from one of the indoor categories for training and the remaining seven are used for validation. Different learning rates are used to perform the experiments and a learning rate of 0.0003 was found to converge well for the training data. The results are summarized in Table 4. Figure 3 shows the images, where the face detection model failed.

4 Discussions

For all the experiments described in the previous section, different learning rates were used for each training set and the best results are presented. In Table 2, different models with different learning rates are trained with MWIR images. The SSD model using MobileNet is the fastest to converge for the data. However, this model is less accurate compared to the other models. When the learning rate is

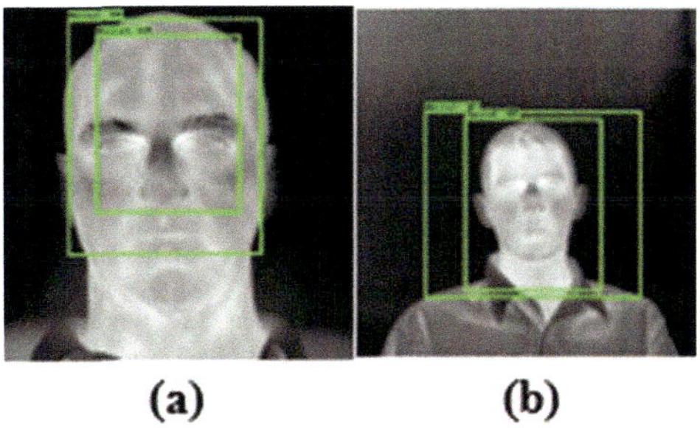

Fig. 3 Failed Detection Outputs – (**a**) when trained on 10 m and validated on 5 m, (**b**) when trained on 5 m and validated on 10 m

further decreased to 3e-6, the model converged after 4 h and the accuracy decreased because of overfitting. R-FCN performed poorly compared to SSD and Faster R-CNN with an accuracy of 91% and took 3.5 h to converge over the training data. It experienced the problem of overfitting when the learning rate is further decreased. Faster R-CNN network converged after training for 2.5 h for the training set with a learning rate of 3e-5 and the accuracy over the validation set is 100%.

An empirical study was conducted to analyze the performance of the Faster R-CNN and to determine when the network starts degrading related to the size of training dataset and the scenarios used for training and testing. It can be seen in Table 3 that, when 80% of the data was used for training, accuracy of the face detection over the validation set was still 100%. The accuracy dropped to 99.26% when 70% of the data was used for training. When the size of the training set is further reduced to 60%, the model did not converge due to a higher learning rate. Learning rate is then decreased to 0.0004 and the accuracy over the test set is 98.72%. This is repeated for 50% and 10% of the images as training data, and the learning rate values used are 0.0004 and 0.001 respectively. The accuracy values are 98% and 96.4% respectively. It can also be noted that the accuracy of the detector is 96.4% when only 10% of the data (622 images) was used for training and the time taken to converge is about 1.5 h. This indicates that this model needs as few as a few hundred images for transfer learning and can perform well when validated using relatively large test dataset (5,600 images in this case). Considering detection speed, determined by the time taken for single detection, SSD with MobileNets is the fastest at 1ms per detection and Faster R-CNN with ResNet-101 is the slowest.

In the next set of experiments, all the images from one of the categories are used for training the model and it is validated over the images from all the other categories one after the other. The learning rate used was 0.0003 for these experiments. The model, when trained on images from 5IN performed well when validated over 5IS (99.87% accurate), however, performance decreased when trained with 5IN alone and validated using 10IN and 10IS. The results were similar when trained with 5IS alone and validated over 5IN, 10IN and 10IS. Accuracy is high when validated over 5IN and dropped down drastically for 10IN and 10IS. When the model is trained with 10IN alone, the accuracy is high when validated over 10IS and the performance degraded when validated over 5IN and 5IS and same is the case for 10IS as training set. It can be inferred from these experiments that, stand-off distance (distance of subject from the camera) had a high impact on the performance of the detector, while applying sunscreen had little impact.

Figure 3 shows two failed cases, the first one is trained using images at 10 m and validated for 5 m, and the second one was trained for 5 m and validated for 10 m. In the second one, though the RPN generated various ROIs, Fast R-CNN network failed to correctly detect the actual position of the face.

5 Conclusions and Future Work

Though there are many algorithms available for face detection, frontal-non/frontal classification, pose estimation, and face recognition in visible spectrum, the algorithms developed or re-trained using thermal images are very limited. In this work, different deep learning models were trained, validated, and optimized using MWIR images collected in different scenarios (different combinations of distances, indoor and outdoor) and the results are compared to determine the one model that works better than others for the task and a suitable learning rate that gave good results for our data.

The models compared are SSD, R-FCN, R-CNN with MobileNet, Inception *v2*, Inception *v3*, ResNet 101 as feature extractors. All the parameters were frozen to be the same as in the original works, only the learning rate was changed to improve the results in terms of accuracy and speed. Faster R-CNN with ResNet-101 performed the best for our data. Also, this model was further trained and tested using different combinations of training and test data, and it performed exceptionally well for training data as small as 10% of the overall data.

Finally, the same model was trained using images from one scenario and validated using images from the other scenarios. The results were poor when the model was trained using images at 5 m and validated using images at 10 m and vice versa. This was due to the fact that the faces in the images at different distances are at different scales as shown in Fig. 1. The efficient model in this case may not be the most effective model for all the object detection applications. It gave good results only for the MWIR images we collected.

However, the data we used for this work is highly limited due to the cost of collection. To overcome this, we plan on scaling the images into different sizes to be able to detect faces in the images at different distances when all we have is only the data collected at one distance. In the future, our efforts will be focused on augmenting the data not only at different scales but also using different contrast settings, and adding different types of noise like Gaussian and filtering out the noises.

This work also presented a frontal/non-frontal classifier with images at 5 m in the indoor setting which was trained and tested only on images that are fully frontal or fully non-frontal (profile). This can be extended to pose estimation (determining the transformation of an object in a 2D image which gives the 3D object) using active shape models.

References

1. Biswas SK, Milanfar P (2017) Linear support tensor machine with LSK channels: pedestrian detection in thermal infrared images. IEEE Trans Image Process 26(9):4229–4242
2. Herrmann C, Ruf M, Beyerer J (2018, April) CNN-based thermal infrared person detection by domain adaptation. In Proceedings of the Autonomous Systems: Sensors, Vehicles, Security and the Internet of Everything, Orlando, FL, USA, pp. 15–19
3. Cross SE, Innes B, Roberts MS, Tsuzuki T, Robertson TA, McCormick P (2007) Human skin penetration of sunscreen nanoparticles: in-vitro assessment of a novel micronized zinc oxide formulation. Skin Pharmacol Physiol 148–154
4. Dai J, Li Y, He K, Sun J (2016) R-FCN: object detection via region-based fully convolutional networks. Adv Neural Inf Proces, vol. abs/1605.06409, 379–387
5. Dowdall J, Pavlidis I, Bebis G (2003) Face detection in the Near-IR spectrum. Image Vis Comput 565–578
6. Eveland CK, Socolinsky DA, Wolff LB (2003) Tracking human faces in infrared video. Image Vis Comput 579–590
7. Farfade SS, Saberian MJ, Li LJ (2015) Multi-view face detection using deep convolutional neural networks. In: Proceedings of the 5th ACM on international conference on multimedia retrieval, pp 643–650
8. Girshick RB (2015) Fast R-CNN. CoRR, pp 91–99
9. He K, Zhang X, Ren S, Sun J (2015) Deep residual learning for image recognition. CoRR, pp 770–778
10. Hochreiter S (1998) The vanishing gradient problem during learning recurrent neural nets and problem solutions. Int J Uncertainty Fuzziness Knowledge Based Syst, 6(2):107–116
11. Howard AG, Zhu M, Chen B, Kalenichenko D, Wang W, Weyand T, Andreett M, Adam H (2017) MobileNets: efficient convolutional neural networks for mobile vision applications. CoRR
12. Huang J, Rathod V, Sun C, Zhu M, Korattikara A, Fathi A, Fischer I, Wojna Z, Song Y, Guadarrama S, Murphy K (2016) Speed/accuracy trade-offs for modern convolutional object detectors. CoRR, pp 7310–7311
13. Jiang H, Learned-Miller E (2017) Face detection with the faster R-CNN. In: 2017 12th IEEE international conference on automatic face gesture recognition (FG 2017), pp 650–657
14. Komatsu S, Markman A, Mahalanobis A, Chen K, Javidi B (2017) Three-dimensional integral imaging and object detection using long-wave infrared imaging. Appl Opt D120–D126
15. Kwaśniewska A, Rumiński J, Rad P (2017) Deep features class activation map for thermal face detection and tracking. In: 2017 10th international conference on human system interactions (HSI). IEEE, pp 41–47
16. Li H, Lin Z, Shen X, Brandt J, Hua G (2015) A convolutional neural network cascade for face detection. In: 2015 IEEE conference on computer vision and pattern recognition (CVPR), pp 5325–5334
17. Liu W, Anguelov D, Erhan D, Szegedy C, Reed S, Fu CY, Berg AC (2016) SSD: single shot multibox detector, pp 21–37
18. Ma C, Trung N, Uchiyama H, Nagahara H, Shimada A, Taniguchi R (2017) Adapting local features for face detection in thermal image. Sensors 2741
19. Ma C, Trung NT, Uchiyama H, Nagahara H, Shimada A, Taniguchi RI (2017) Mixed features for face detection in thermal image. Proc SPIE Int Soc Opt Eng 103380E
20. Murata T, Matsuno S, Mito K, Itakura N, Mizuno T (2017) Investigation of facial region extraction algorithm focusing on temperature distribution characteristics of facial thermal images. In: HCI international 2017 – posters' extended abstracts, pp 347–352
21. Ranjan R, Patel VM, Chellappa R (2017) HyperFace: a deep multi-task learning framework for face detection, landmark localization, pose estimation, and gender recognition. IEEE Trans Pattern Anal Mach Intell, Vol. abs/1603.01249, 121–135

22. Reese K, Zheng Y, Elmaghraby AS (2012) A comparison of face detection algorithms in visible and thermal spectrums. In: Object recognition supported by user interaction for service robots
23. Ren S, He K, Girshick RB, Sun J (2015) Faster R-CNN: towards real-time object detection with region proposal networks. CoRR, pp 91–99
24. Sun X, Wu P, Hoi SCH (2017) Face detection using deep learning: an improved faster R-CNN approach. Neurocomputing 42–50
25. Szegedy C, Liu W, Jia Y, Sermanet P, Reed S, Anguelov D, Erhan D, Vanhoucke V, Rabinovich A (2014) Going deeper with convolutions. In: Proceedings of the IEEE conference on computer vision and pattern recognition, pp 1–9
26. Szegedy C, Vanhoucke V, Ioffe S, Shlens J, Wojna Z (2015) Rethinking the inception architecture for computer vision. CoRR, pp 2818–2826
27. Yang S, Luo P, Loy CC, Tang X (2015) From facial parts responses to face detection: a deep learning approach. In: 2015 IEEE international conference on computer vision (ICCV), pp 3676–3684
28. Yang S, Luo P, Loy CC, Tang X (2018) Faceness-Net: face detection through deep facial part responses. IEEE Trans Pattern Anal Mach Intell, Vol. abs/1701.08393, 1845–1859
29. Yang S, Xiong Y, Loy CC, Tang X (2017) Face detection through scale-friendly deep convolutional networks. CoRR
30. Zhang K, Zhang Z, Li Z, Qiao Y (2016) Joint face detection and alignment using multitask cascaded convolutional networks. IEEE Signal Process Lett, vol. abs/1604.02878, 1499–1503
31. Zheng Y (2012) Face detection and eyeglasses detection for thermal face recognition. 83000C
32. Zhu C, Zheng Y, Luu K, Savvides M (2017) CMS-RCNN: contextual multi-scale region-based CNN for unconstrained face detection, pp 57–79

Part III
Mobile-Based Active Authentication

Mobile Active Authentication based on Multiple Biometric and Behavioral Patterns

Alejandro Acien, Aythami Morales, Ruben Vera-Rodriguez, and Julian Fierrez

Abstract In this chapter we evaluate mobile active authentication based on an ensemble of biometrics and behavior-based profiling signals. We consider seven different data channels and their combination. Touch dynamics (touch gestures and keystroking), accelerometer, gyroscope, WiFi, GPS location and app usage are all collected during human-mobile interaction to authenticate the users. We evaluate two approaches: one-time authentication and active authentication. In one-time authentication, we employ the information of all channels available during one session. For active authentication we take advantage of mobile user behavior across multiple sessions by updating a confidence value of the authentication score. Our experiments are conducted on the semi-uncontrolled UMDAA-02 database. This database comprises of smartphone sensor signals acquired during natural human-mobile interaction. Our results show that different traits can be complementary in terms of mobile user authentication and multimodal systems clearly increase the performance when compared to individual biometrics systems with accuracies ranging from 82.2% to 98.0% depending on the authentication scenario.

Keywords Mobile authentication · Multimodal approaches · Behavioral patterns · Behavioral-based profiling · Touch dynamics

The present chapter is adapted from the conference paper A. Acien et al. "MultiLock: Mobile Active Authentication based on Multiple Biometric and Behavioral Patterns", in ACM Intl. Conf. on Multimedia, Workshop on Multimodal Understanding and Learning for Embodied Applications (MULEA), pp. 53–59, Nice, France, October 2019. The new material here includes Table I and Sect. 5.3.

A. Acien (✉) · A. Morales · R. Vera-Rodriguez · J. Fierrez
Biometrics and Data Pattern Analytics (BiDA) Lab, EPS, Universidad Autonoma de Madrid, Madrid, Spain
e-mail: alejandro.acien@uam.es; aythami.morales@uam.es; ruben.vera@uam.es; julian.fierrez@uam.es

T. Bourlai et al. (eds.), *Securing Social Identity in Mobile Platforms*, Advanced Sciences and Technologies for Security Applications,
https://doi.org/10.1007/978-3-030-39489-9_9

1 Introduction

Services are migrating from the physical to the digital domain in the information society. Examples can be found in e-government, banking, education, health, commerce, and leisure. This digital revolution is associated with a massive deployment of mobile devices, including multiple sensors (e.g. camera, gyroscope, GPS, touch screens, etc.), and full connectivity (e.g. bluetooth, WiFi, 4G, etc.). The mobile market has expanded to the point where the number of mobile devices in use is nearly equal to the world's population [1]. Mobile devices are rapidly becoming data hubs, used to store e-mail, personal photos, online history, passwords, and even payment information. Recent studies have shown that about 34% or more users do not use any form of authentication mechanism on their devices [2]. In similar studies, inconvenience is always shown to be one of the main reasons why users do not use any authentication mechanism. In [3], researchers show that mobile device users spent up to 9% of the time they use their smartphone on unlocking their screens, and the 2018 Meeker Report [4] indicated that the average smartphone user checks his/her device 150 times per day. Those factors lead individuals to make less security conscious decisions like leaving their smartphones unprotected or just protecting them using simple to break authentication mechanisms (e.g., simple Google unlock graphical patterns vulnerable to over-the-shoulder attacks [5]).

Biometric technologies improve traditional recognition technologies in several ways based on passwords or swipe patterns. The advantages of biometric systems are many in terms of security and convenience of use, which has led these technologies to take on a leading role in the last years. In fact, there is a growing interest in the biometrics research community towards more transparent and robust authentication methods that make use of the interaction signals originated when using smartphones [6, 7]. Signals generated with the sensors already embedded in mobile devices (e.g., gyroscope, magnetometer, accelerometer, GPS, and touchscreen interactions) along with metadata associated to our use of the technology (e.g. internet point access, browsing history, app usage) could assist in user authentication avoiding the inconveniences of traditional unlocking systems. All this information is originated naturally during the normal usage of the user with a smartphone, and it has been demonstrated that can be used for person identification under certain conditions [7]. By regularly conducting unobtrusive identity checks of the mobile user during a normal session, a continuous authentication system can verify if the device is still being operated by the authorized user. With this active system, if the mobile device is stolen, it should quickly recognize the presence of an unauthorized user.

The aim of this chapter is to analyze multi-modal approaches to improve the performance of mobile authentication. Our experiments include up to four different biometric traits (touch gestures, keystroking, gyroscope, and accelerometer) and three behavioral-based profiling techniques (GPS, WiFi, and app usage). The experiments are conducted on the UMDAA-02 mobile database [8], a challenging dataset acquired under natural conditions.

Previous works have demonstrated the potential of biometric and behavioral-based profiling patterns for user authentication under controlled scenarios. However, the performance of biometric mobile authentication based on human interaction raises doubt under challenging non-supervised scenarios. The contributions of this work are: (i) performance analysis of user authentication based on 4 biometric data channels (touch gestures, keystroking, accelerometer, and gyroscope) and 3 behavior profiling data sources (WiFi, GPS, and App usage), obtained during natural human-smartphone interaction; (ii) study of multimodal approaches for smartphone user authentication based on various combinations of the previous 7 data channels, both for One-Time Authentication and for Active Authentication schemes (i.e., continuously over multiple sessions). The results showed in this chapter suggest that user-profiling techniques can help to improve performances of behavioral-based biometrics authentication systems in all scenarios evaluated.

The rest of this chapter is organized as follows: Section 2 links the present works with related research. Section 3 describes the architecture of our approach. Section 4 explains the experimental protocol, describing the database and the experiments performed. Section 5 presents the final results for single and multimodal architecture and Sect. 6 summarizes the conclusions and future work.

2 Related Works

Mobile authentication based on soft biometrics traits has been extensively studied in the last years [9–11]. In Table 1 we summarize some of the most relevant state-of-the-art works in this field. Swipe dynamics is one of the most popular traits analyzed [9]; however, it has been shown not to have enough discriminative power to replace traditional technologies.

Accelerometer and gyroscope sensors have been studied traditionally for gait recognition, and some works have demonstrated also their utility for user authentication with acceptable performance [12].

Geo-location based verification approaches are scarce in the literature. In [13], Mahbub and Chellappa developed a mobile authentication system using trace histories by generating a confidence score of the new user location taking into account the sparseness of the geo-location data and past locations. For this purpose, they employed modified Hidden Markov Models (HMMs) considering the human mobility as a Markovian motion. In a similar way, in [14] a variation of HMMs was used to develop a user authentication mobile system by exploiting application usage data. They suggest that unforeseen events and unkno applications have more impact in the authentication performance than the most common apps used by the user. The potential of WiFi history data was analyzed in [10] for mobile authentication. They explored: (i) the WiFi networks detected by the smartphone, (ii) when the detection occurs, and (iii) how frequently those networks are detected during a period of time.

Regarding keystroke traits, in [11] a fixed-text keystroking system for mobile user authentication was studied using not only time and space based features (e.g.

Table 1 Summary of the state-of-the-art in biometric mobile authentication. The number of users for each database is in brackets

Study	Modality	Classifier	Database	Acc (%)
Fierrez et al. [9]	Touch gestures	SVM, UBM	Serwadda (190), Antal (71), Frank (41), UMDAA02 (48)	80–90
Li et al. [10]	Accelerometer, WiFi	Templates, random Forest	Prop. DB (321)	90 (3 s)
Busheck et al. [11]	Keystroke	KNN, SVM, NB, LSAD	Prop. DB (28)	64–74
Li et al. [12]	Accelerometer, gyroscope	Random Forest	Prop. DB (304)	77
Mahbub et al. [13]	GPS location	M-HMM	UMDAA02 (48), GeoLife (182)	69–79
Mahbub et al. [14]	App usage	M-HMM	UMDAA02 (48)	70–84
Monaco et al. [15]	Keystroke	POHMM, HMM, SVM	Mutiple Databases (247)	90
Shi et al. [18]	Voice, GPS, touch, gait	NB	Prop. DB	90
Fridman et al. [19]	Stylometry, app usage, web browsing, GPS	Binary classifiers, SVM	Prop. DB (200)	95 (3 s)
Liu et al. [20]	Touch gestures, power consumption, accelerometer, gyroscope, magnetometer	SVM	Prop. DB (10)	95
Li et al. [21]	WiFi, Bluetooth, accelerometer, gyroscope	Random Forest	Prop. DB (321)	90 (3 s)
Deb et al. [22]	Keystroke, GPS, accelerometer, gyroscope, magnetometer, accelerometer, gravity, rotation sensors	Siamese LSTM	Prop. DB (37)	97 (3 s)
Our work	Touch gestures, keystroke, accelerometer, gyroscope, WiFi, GPS, app usage	SVM, templates	UMDAA02 (48)	98 (4 sessions)

hold and flight times, jump angle or drag distance) but also studying the hands postures during typing as discriminative information. In [15], a novel fixed-text authentication system for laptops and mobile devices based on Partially Observable HMMs was studied. This model is an extension of HMMs, in which the hidden state is conditioned on an independent Markov chain. The algorithm is motivated by the idea that typing events depend both on past events and also on a separate process.

Finally, building a multimodal system that integrates all these heterogeneous information sources for mobile user authentication is still a challenge [16]. Noisy data, intra class variation or spoofing attacks [17] are some inevitable problems in unimodal systems that can be overcome by multimodal architectures [7, 16]. In [18], a multimodal user authentication system was based on the fusion at decision level of voice, location, multi-touch, and accelerometer data. Their preliminary results suggest that these four modalities are suitable for continuous authentication. In [19], a fusion was performed also at decision level of behavioral-based profiling signals such as web browsing, application usage, and GPS location with keystroking data achieving 95% of user authentication accuracy using information from one-minute window.

More recently, in [20] a fusion also at decision level of touch dynamics, power consumption, and physical movements modalities achieved 94.5% of accuracy with a dataset that was captured under supervised conditions. In [21], an unobtrusive mobile authentication application is designed for single and multimodal approaches. They collected data from WiFi, Bluetooth, accelerometer, and gyroscope sources in unsupervised conditions and fused them at score level achieving up to 90% of accuracy in the best scenario. In [22], they propose a Siamese Long ShortTerm Memory network architecture to merge up to 8 modalities (keystroke dynamics, GPS location, accelerometer, gyroscope, magnetometer, linear accelerometer, gravity, and rotation sensors) for mobile authentication, achieving 97.15% of accuracy using data from a 3 s window for each of the modalities considered individually.

Previous works fusing different modalities ([19, 21, 22]) have focused their approach on obtaining time windows from the different modalities and then carry out the fusion. However, this does not represent a realistic scenario due to not all modalities fused can always be captured in a specific time windows. In this work we go a step forward by merging the modalities at session level (time during an unlock and the next lock of the device), and therefore fusing only the modalities available at each session.

3 System Description

In this chapter we analyze 4 biometric data channels (touch gestures, keystroking, gyroscope, and accelerometer) and 3 behavior data sources (GPS, WiFi, and app usage). We study 2 approaches for user authentication (see Fig. 1):

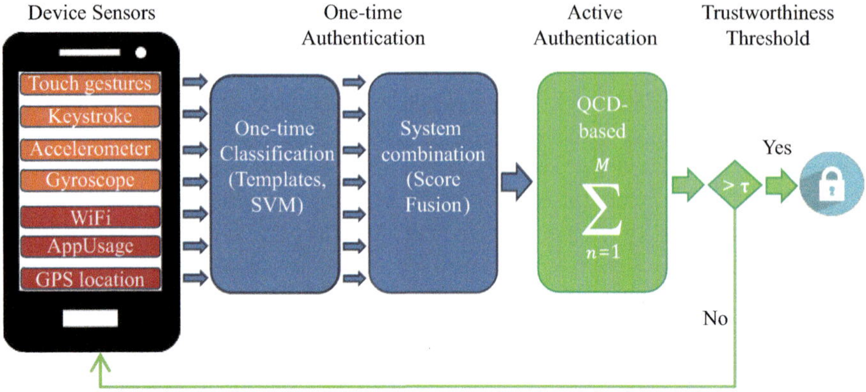

Fig. 1 System architecture. Blue boxes correspond to one-time authentication, and green boxes are add-on modules for active authentication

- The first approach (blue boxes in Fig. 1), referred to as One-Time Authentication (OTA) is based on unimodal systems trained with the information extracted from the mobile sensors during a user session. Remember that a session is defined as the elapsed period between the device unlock and the next lock. Therefore, sessions have a variable duration and information obtained from sensors varies depending on the usage of the device during the session. The information provided by the sensors is employed to model the user according to the seven systems mentioned before: keystroking, touch gestures, accelerometer, gyroscope, WiFi, app usage, and GPS location. Each system provides a single authentication score and these scores are combined to generate a unique score for each session.
- The second approach, called Active Authentication (green boxes in Fig. 1), is based on updating a confidence value generated from the One-Time Authentication during consecutive sessions.

The seven systems are categorized into two main groups according to the nature of the information employed to model the user: biometric and behavior-based profiling systems. In this work, biometric systems refer to the top 4 channels in the Sensors Data module of Fig. 1 (red boxes). The way we carry out touch gestures, typing, or handle the device is determined by behavioral aspects (e.g. emotional state, attention) and neuromotor characteristics of users (e.g. ergonomic, muscles activation/deactivation timing, motor abilities). Behavioral-based profiling refers to those systems that identify the owners of the device according to the services they use during their daily habits (orange boxes in Fig. 1, bottom 3 channels in the Sensors Data module).

Table 2 Example of an app-usage user template generated according the data captured during 6 days

Event	Time slot	Frequency
WhatsApp	4	5
Navigator	4	3
YouTube	5	1
WhatsApp	5	1
Facebook	7	2

3.1 Behavioral-Based Profiling Systems

WiFi, app usage, and GPS location system are based on a similar template-based matching algorithm. A user template is defined as a table containing the time stamps and the frequency of the events [10]. For this, we divided the time (24 h of the day) into N equal time slots (e.g. if we choose N = 48 we will have 48 time slots of 30 min), giving to each time slot a number ID. Then, we store in the template the event's name, the number ID of the time slot and the occurrence frequency of that event (number of times this event occurs during this particular time slot on a window of consecutive days). Table 2 shows an example of the app-usage template for a given user generated according the data obtained during 6 days; in this case the WhatsApp application, for instance, is detected in the fourth time slot during five out of six total days considered, meanwhile the same app is detected only one out of 6 days in the fifth time slot. Note that multiple detections of the same event in the same time slot and day are ignored, but they are stored if they belong to different time slots or days. Depending on the system, the event could be the name of the WiFi network, latitude and longitude of a location (with two decimals of accuracy), or the name of a mobile app for WiFi, GPS location, and app usage systems, respectively.

Finally, we test the systems by calculating a behavior-based confidence score [10] for each test session as:

$$score = \sum\nolimits_{i=1}^{S} f_i^2 \tag{1}$$

where *fi* is the frequency of the event stored in the template that match with the test event i in the same time slot and S is the total number of events detected in that test session. For example, if the test session includes the usage of *WhatsApp* and *Navigator* apps during the fourth slot, the score confidence will be 52 + 32 = 34 (according to the template showed in Table 1). Based on this explanation, a higher score in the test session implies higher confidence for authentication.

3.2 Biometric Systems

For touch gestures, keystroking, accelerometer and gyroscope systems, the feature extraction and classification algorithms are adapted to model the user information.

In the touch gestures system, the feature set employed is a reduced set of the global features presented in [23] (commonly used for online handwriting sequence modeling) and adapted for swipe biometrics in [7]. Mean velocity, max acceleration, distance between adjacent points, or total duration are some examples of this subset of 28 features extracted (see [23] for details).

For accelerometer and gyroscope, the data captured are comprised of the x, y, and z coordinates of the inclination vector of the device (gyroscope) and the acceleration vector (accelerometer) in each time stamp. For these 2 sensors we use the feature set proposed in [12]: mean, median, maximum, minimum, distance between maximum and minimum, and the standard deviation for each array of coordinates. Moreover, we propose the 1 and 99 percentiles[1] and the distance between them as additional features.

Regarding keystrok dynamics, the keys pressed were encrypted in order to ensure users' privacy. Thus, systems based on graphs were discarded and we adopted traditional timing features: hold time, press-press latency, and press-release latency as in [24, 25]. Finally, we propose a feature set based on six statics (mean, median, standard deviation, 1 percentile, 99 percentile, and 99-1 percentile). Note that UMDAA-02 keystroke data can be considered as a free text scenario. However, the limited samplesper session and the encrypted keys difficult the application of popular free-text keystroke authentication methods.

For classification we train different Support Vector Machines (SVM) with a radial basis function (RBF) kernel, one for each feature set and user with an optimization of both hyperparameters (C, σ).

4 Experiments

4.1 Database

The experiments were conducted with UMDAA-02 database [8]. This database comprises 141.14 GB of smartphone sensor signals collected from 48 Maryland University students over a period of 2 months. The participants used a smartphone provided by the researchers as their primary device during their daily life (unsupervised scenario). The sensors captured are touchscreen (i.e. touch gestures and keystroking), gyroscope, accelerometer, magnetometer, light sensor, GPS, and WiFi, among others. Information related to mobile user's behavior such as lock and unlock time events, start and end time stamps of calls and app usage are also stored. Table 3 summarizes the characteristics of the database. During a session, the data collection application stored the information provided by the sensors in use.

[1] Indicate the value below which a given percentage of observation (samples in this case) in a group of observation falls.

Table 3 General UMDAA-02 dataset information

Description	Statistics
Gender	36 M/12F
Age	22–31 years
Avg. days/user	10 days
Avg. sessions/user	248 sessions
Avg. time/session	224 s
Avg. systems/session	5.2 systems[a]

[a]Systems: refers to the number of systems available out of the 7 studied in this work

4.2 Experimental Protocol

The experiments are divided into two different scenarios: One-Time Authentication (OTA) and Active Authentication (AA). In OTA the performance is calculated using only one session to authenticate the user meanwhile in AA we employ multiple consecutive sessions in order to improve the confidence in the authentication. For all experiments the dataset is divided into 60% days for training (first sessions) and the remaining 40% days for testing in order to have train and test sets as more balanced as posible. This means that we employ 6 days in average to model the user and 4 days in average to test such a model. The performance for both scenarios is presented in terms of average correct classification rate computed as $100 - EER$ (Equal Error Rate).[2]

4.2.1 One-Time Authentication

In OTA experiments, all 7 systems are trained separately for each user and the scores are calculated at session level, generating 7 scores for each test session as maximum (note that the number of systems available during a session varies). The 4 biometric systems considered can produce more than one score per session (e.g. multiple gestures or multiple keystroking sequences during a text chat). In those cases, the scores available during the session are averaged to obtain one score for each biometric system and session. Finally, we normalize with *tan* normalization and fuse the scores (mean rule) to calculate a single score [14] according to the different fusion set-ups proposed. The scores from the best fusion set-up will be used in the AA scenario (details are provided in Sect. 4.2.2 below).

[2]EER refers to the value where False Acceptance Rate (percentage of impostors classified as genuine) and False Rejection Rate (percentage of genuine users classified as impostors) are equal.

4.2.2 Active Authentication

For AA experiments we consider the QCD algorithm (Quickest Change Detection) as explained in [26]. The QCD-based algorithm updates a confidence score based on previous events (sessions in this work) by performing a cumulative sum of scores. This cumulative sum will be almost zero if the scores belong to the genuine user, and will grow if an impostor takes the control, until it reaches a certain threshold that would detect the intruder. The cumulative sum is calculated as follow:

$$score_j^{AA} = \max\left(score_{j-1}^{AA} + L_j, 0\right) \quad (2)$$

where j means the actual session and $score_{j-1}^{AA}$ is the previous cumulative score. L_j is the contribution of the actual session calculated as the log-likelihood ratio between score distributions:

$$L_j = \log\left(\frac{f_I\left(score_j\right)}{f_G\left(score_j\right)}\right) \quad (3)$$

where f_G and f_I are the probability distributions of the genuine and impostor scores respectively calculated previously in the OTA fusion scenario, and *scorej* is the OTA fused score of the actual session. According to (3), the log-likelihood ratio L_j will be negative if $score_j$ belongs to a genuine user and positive in the opposite case and, therefore, multiple consecutive sessions of an impostor in control will increase the cumulative sum ($score_j^{AA}$). Figure 2 depicts an example of $score_j^{AA}$ evolution. At the time the mobile starts to be operated by an intruder (session number sixteen in Fig. 2) the $score_j^{AA}(j > 16)$ will tend to increase until reaching the threshold. The time elapsed between the intrusion start and the intrusion detection is known as Detection Delay (DD) measured in number of sessions.

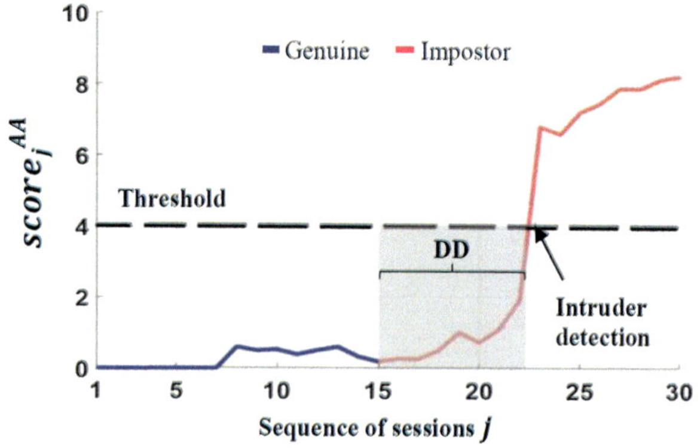

Fig. 2 An example of QCD-based curve with a sequence of 30 sessions (15 genuine and 15 impostors). The dashed line is the intruder detection threshold and the grey box shows the Detection Delay (DD)

5 Results and Discussion

5.1 One-Time Authentication

In this section we analyze the OTA scenario: the accuracy for the 4 biometric systems and the fusion with each behavior-based profiling system. Table 4 summarizes the final results by ranking from the best individual biometric system performance to the worst one. The first column shows the performance obtained for each single biometric system. From the second to the fourth column, we show the performance for the fusion of each biometric system with each behavior-based profiling system, and the fifth column shows the fusion with all of them. Firstly, the poor performance achieved by some biometric systems can be caused by the uncontrolled acquisition conditions of the database and the limited number of samples per session (e.g. free text keystroke usually requires large sequences) but the combination of all of them (last row in Table 4) shows acceptable performance for unsupervised scenarios. Secondly, we can observe that behavior-based profiling systems always improve biometric systems performances in all fusion schemes. In fact, the combination of all behavior-based profiling approaches with each biometric system achieves the most competitive performance, improving them in more than 18% of accuracy in the best case. If we analyze each single behavior-based profiling fusion, we can observe that the GPS system achieves the best improvements, boosting biometric systems performances in more than 13% of accuracy. Finally, in Fig. 3 we plot the ROC curves for each single biometric system and the best fusion set-up, i.e. the fusion of all behavior-based profiling systems with each biometric system (column 5 in Table 4). The results in OTA scenario suggest that behavior-based profiling systems always improve the biometric ones and the best performance is achieved by fusing with all of them, and therefore, the scores obtained from this fusion scheme will be used in the AA scenario.

Table 4 Results achieved for both One-Time and Active Authentication (AA) scenarios in terms of correct classification rate (%) according to different number of biometric systems and their fusion with behavior-based profiling systems. In brackets, average number of sessions employed (ADD)

System	Acc.	+WiFi	+ GPS	+ App usage	All	AA
Touch gestures	72.0	78.2	78.3	75.4	**83.1**	95.0 (6)
Keystroking	62.5	72.6	70.9	67.8	**79.1**	92.9 (7)
Accelerometer	61.3	70.8	77.3	64.7	**78.7**	93.7 (7)
Gyroscope	59.5	69.7	72.6	63.4	**78.4**	92.3 (6)
Combined	73.2	77.3	78.9	75.3	**82.2**	**97.1 (5)**

Bold indicates the best accuracy achieved for each system

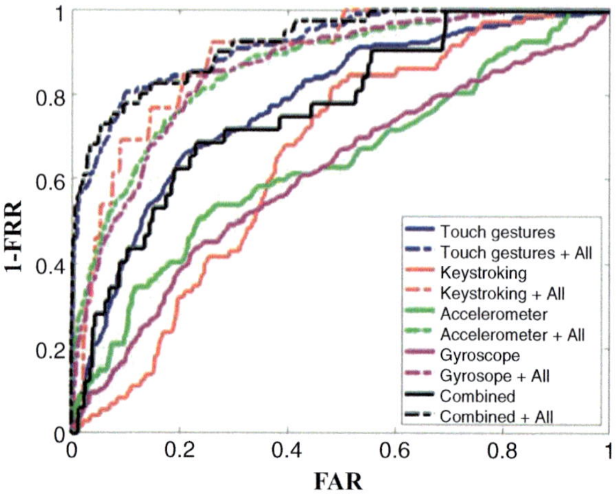

Fig. 3 ROC curves (One-Time Authentication) for individual biometrics and the best fusion set-up incorporating the 3 considered behaviour profiling sources (All = WiFi + GPS + App usage)

5.2 Active Authentication

Even performance metrics used for Active Authentication and One-time Authentication can be similar, we want to highlight some important differences:

- *Probability of False Detections (PFD):* is the percentage of genuine users detected as intruder during a sequence of genuine sessions. It means that $score_j^{AA}$ reaches the intruder detection threshold during a genuine session sequence (genuine curve in Fig. 2). PFD is similar to FMR (False Match Rate) in One-Time Authentication.
- *Probability of Non-Detection (PND):* is the percentage of intruders not detected during a sequence of intruder sessions. It means that $score_j^{AA}$ does not reach the intruder detection threshold during the intruder sessions sequence (impostor curve in Fig. 2). PND is similar to FNMR (False Non-Match rate) in One-Time Authentication.
- *Average Detection Delay (ADD):* is the average number of impostor sessions needed to detect an intruder (the grey box in Fig. 2).

To calculate the correct classification rate in AA we plot in Fig. 4 the PND vs. PFD and ADD vs. PFD curves. The PND-PFD curves are similar to FMR-FNMR curve in one-time authentication with the main difference that those results are obtained from a sequence of stacked scores instead of only one. The equal error rate (EER) will be the value where PND and PFD are equal and the correct classification rate will be computed as $100 - EER$. The ADD-PFD curve shows the number of sessions needed to detect an intruder according to the PFD. This curve allows us to know how many sessions are needed to achieve the EER reported. For instance, the PND-PFD curves in Fig. 4 (right) show that the EER in Active Authentication is 2.9% and the ADD to achieve that EER is 5 sessions. This means that we can

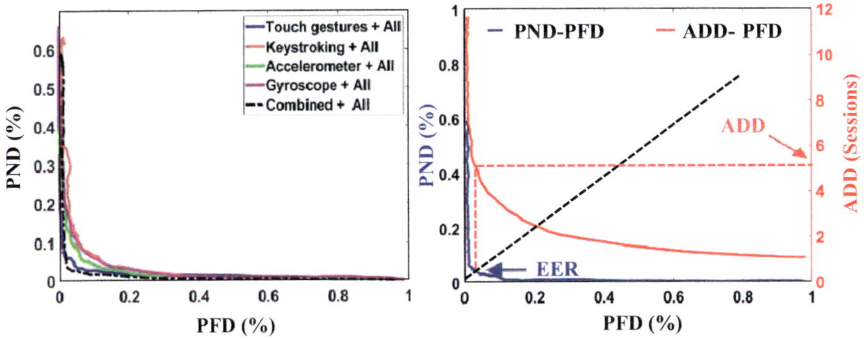

Fig. 4 PND vs PFD curves of active authentication for the best fusion schemes (left), PND vs PFD and ADD vs PFD curves for the best fusion set-up (right). The dark dashed line shows the EER and the red one shows the Average Detection Delay for that EER in the right plot

improve OTA results at the cost of having more sessions to detect an intruder. All curves were calculated for each user and averaged.

Finally, all AA results are summarized in the last column of Table 4. Remember that scores employed in the QCD-based algorithm come from the fusion scores of the best OTA scenario (fusing with all behavior-based profiling systems) so both performances are correlated. Each performance in Table 4 for AA is followed by the average detection delay in brackets needed to achieve it. As we expected, in all different fusion set-ups the AA algorithm improves the accuracy at the cost of needing more sessions to detect the intruder. In fact, for the best fusion set-up the performance improves from 82.2% to 97.1% by using 5 consecutive intruder sessions to detect the impostor. Comparing all scenarios, the greatest improvement occurs with all biometric systems combined (14.9% of improvement in the last row of Table 4) with an average 5 sessions.

The cost of need up to 5 sessions to detect an intruder could be unacceptable in some real life scenarios (e.g. prevent unauthorized use of devices during distractions). However, as some recently surveys suggest [27], the market of secondhand mobile phone is constantly growing and some of these devices have a provenance of dubious legality. In these scenarios Active Authentication approaches can serve to persuade burglars and unathorized usages.

5.3 *Temporal Dependency in Behavioral-Based Profiling Systems*

As mentioned in [10], the performance of behavioral-based profiling systems could be affected by differences in our routines in our daily life. For example, the places we visit during the week could vary at weekends or the WiFi signals detected are different if we are at work (working hours) or at home (leisure hours). In order to

Table 5 Results achieved for behavioral-based profiling systems and the Combined + All fusion scenario according to the temporally division in week/weekend and working/leisure time

Profiling System	Acc.	Week	Weekend	Working time	Leisure time
WiFi	77.5	77.1	77.9	74.4	**85.0**
GPS	75.4	74.0	80.1	70.1	**83.7**
App usage	67.4	67.6	69.2	66.2	**69.7**
Combined + all	82.2	81.6	82.0	82.1	**86.7**

study these assumptions, we divide the score sessions of the OTA scenario in two groups depending on when the session was performed: week time (from Monday to Friday) or weekend (Saturday and Sunday), and working time (from 9 a.m to 6 p.m from Monday to Friday) or leisure time (the remaining hours for all days of the week). The results are shown in Table 5 for all behavioral-based profiling systems separately (i.e. using only the scores of profiling systems in each sessions) and the best fusion scenario of OTA (Combined + All).

First of all, the results of week/weekend division suggest that only GPS system improves their performance by using the scores of the weekend sessions due to at weekends the users move to more locations than during the week, making the GPS system more discriminant at weekends. Secondly, if we divide the sessions according to whether they belong to working time or leisure time we find out that all behavioral-based profiling systems improve their performance in leisure hours, specially GPS and WiFi systems. The invariance across daily hours of App usage system suggest that users usually employ almost the same mobile applications at work as they do at home.

It is worth noting that these conclusions relate heavily on the nature of the database considered, where most of the users are students of the same university, who therefore share similar location patterns during weekdays and working hours. On a much broader database we would probably expect to have different trends, maybe having better results for working hours where probably users follow a more constant location pattern.

Finally, the last row in Table 5 shows the variation of the best fusion set-up in OTA scenario according to the proposed time division. As we expected, the best improvement is achieved during leisure time (see Fig. 5 for details) due to the improvement of the behavioral-based profiling systems, raising the accuracy up to 86.7%.

Regarding AA scenario, if we employ only the sessions from leisure time in the best fusion OTA set-up (combined + All), we achieve an accuracy of 98.0% with 4 sessions, improving the best result of Table 4, where we achieved a 97.1% of accuracy for AA with 5 sessions, but considering data from all hours and not only leisure time.

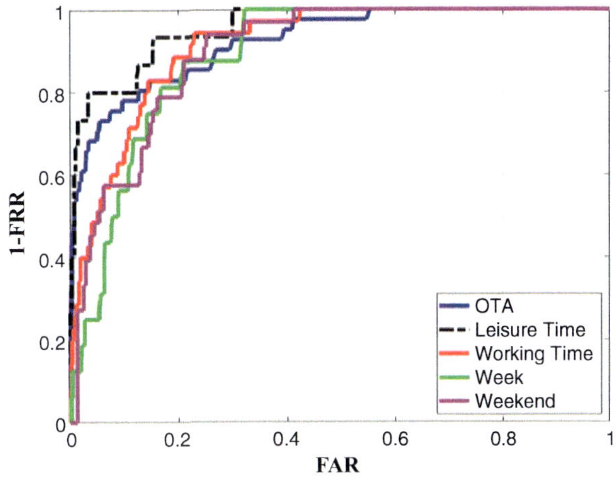

Fig. 5 ROC curves for the best OTA scenario (combined + ALL) according to the proposed time division: week/weekend, working/leisure time

6 Conclusions and Future Work

In this chapter, we have studied user mobile active authentication based on multiple biometric and behavior-based profiling systems. For this, we studied two scenarios according to the number of sessions used: one session (One-Time Authentication) and multiple sessions (Active Authentication). The results suggest that some swipe and keystroking modalities work better than accelerometer and gyroscope in the scenarios evaluated in this work. The fusion with behavior-based profiling systems improves the results of single biometric modalities, achieving accuracies up to 82.2% in the best case for an OTA scenario. Our experiments also suggest that Active Authentication improves the accuracy of One-time Authentication scenario with up to 14% of enhancement using information from 5 sessions. As we mentioned in the section before, Active Authentication algorithms are useful in those scenarios where the intruder attempt to use the mobile phone during mid-term periods (e.g. to use it as their personal device, reselling in the second-hand mobilephone market, etc). According to this, a continuous usage of the stolen mobilephone in which One-Time verification system has already been hacked and intruders has no limitations regarding device's usage. Active Authentication continuously monitorizes and check the identity of users. These approaches can serve to persuade robberies and unauthorized usages.

For future works we will work to improve the performance of individual systems, especially biometrics systems. Better individual performances will produce better fused schemes. The combination of heterogeneous data at data and feature level will be evaluated in order to merge correlations between systems (e.g. touch gestures and apps used are highly correlated).

Regarding the temporary dependency in behavioral-based profiling systems, note that all participants of UMDAA02 database are students from the same university so probably some of them share work places and leisure activities, and therefore, these variations reported could be greater in other mobile databases with users from different places and habits.

Acknowledgments This work was funded by the projects BIBECA (RTI2018-101248-B-I00 MINECO/FEDER) and Bio-Guard (Ayudas Fundación BBVA a Equipos de Investigación Científica 2017), and by CECABANK.

References

1. Radicati S (2018) Mobile statistics report, 2014–2018. The Radicati Group, INC. A Techonology Market Research Firm, Palo Alto
2. Cho G, Huh JH, Cho J, Oh S, Song Y, Kim H (2017) SysPal: system-guided pattern locks for android. In: Proceedings of IEEE Symposium on Security and Privacy, California, UE
3. Harbach M, von Zezschwitz E, Fichtner A, Luca AD, Smith M (2014) It's a hard lock life: a field study of smartphone (un)locking behavior and risk perception. In: Proceedings of symposium on usable privacy and security, California, USA
4. Molla R (2018) Mary Meeker's 2018 internet trends report: all the slides, plus analysis. In Recode
5. Martinez-Diaz M, Fierrez J, Galbally J (2016) Graphical password-based user authentication with free-form doodles. IEEE Trans Human-Machine Syst 46(4):607–661
6. Crouse D, Han H, Chandra D, Barbello B, Jain AK (2015) Continuous authentication of Mobile user: fusion of face image and inertial measurement unit data. In: Proceedings of IAPR international conference on biometrics, Phuket, Thailand
7. Patel VM, Chellappa R, Chandra D, Barbello B (2016) Continuous user authentication on mobile devices: recent progress and remaining challenges. IEEE Signal Process Mag 33(4):49–61
8. Mahbub U, Sarkar S, Patel VM, Chellappa R (2016) Active user authentication for smartphones: a challenge data set and benchmark results. In: Proceedings of IEEE 8th international conference on biometrics theory, applications and systems, New York, USA
9. Fierrez J, Pozo A, Martinez-Diaz M, Galbally J, Morales A (2018) Benchmarking touchscreen biometrics for Mobile authentication. IEEE Trans Inf Forensics Sec 13(11):2720–2733
10. G. Li and P. Bours (2018). Studying WiFi and accelerometer data based authentication method on mobile phones. In: Proceedings of 2nd international conference on biometric engineering and applications, Amsterdam, Netherlands
11. Buschek D, De Luca A, Alt F (2015) Improving accuracy, applicability and usability of keystroke biometrics on mobile touchscreen devices. In: Proceedings of 33rd annual ACM conference on human factors in computing systems, Seoul, Republic of Korea
12. Li G, Bours P (2018) A novel mobilephone application authentication approach based on accelerometer and gyroscope data. In: Proceedings of 17th international conference of the biometrics specials interest group, Fraunhofer, Germany
13. Mahbub U, Chellappa R (2016) PATH: person authentication using trace histories. In: Proceedings of ubiquitous computing, electronics & mobile communication conference, IEEE, New York, USA
14. Mahbub U, Komulainen J, Ferreira D, Chellappa R (2018) Continuous authentication of smartphones based on application usage. IEEE Transactions on Biometrics, Behavior, and Identity Science 1(3):165–180
15. Monaco JV, Tappert CC (2018) The partially observable hidden Markov model and its application to keystroke dynamics. Pattern Recogn 76:449–462

16. Fierrez J, Morales A, Vera-Rodriguez R, Camacho D (2018) Multiple classifiers in biometrics. Part 2: trends and challenges. Inf Fusion 44:103–112
17. Marcel S, Nixon MS, Fierrez J, Evans N (2019) Handbook of biometric anti-spoofing. presentation attack detection, Advances in computer vision and pattern recognition. Springer, Cham
18. Shi W, Yang J, Jiang Y, Yang F, Xiong Y (2011) Senguard: passive user identification on smartphones using multiple sensors. In: Proceedings of 7th IEEE international conference on wireless and mobile computing, networking and communications, Shangai, China, pp 141–148
19. Fridman L, Weber S, Greenstadt R, Kam M (2015) Active authentication on mobile devices via stylometry, GPS location, web browsing behavior, and application usage patterns. IEEE Syst J 11(2):513–521
20. Liu X, Shen C, Chen Y (2018) Multi-source interactive behavior analysis for continuous user authentication on smartphones. In: Proceedings of Chinese conference on biometric recognition, Urumchi, China
21. Li G, Bours P (2018) A mobile app authentication approach by fusing the scores from multi-modal data. In: Proceedings of 21st international conference on information fusion, Cambridge, UK
22. Deb D, Ross A, Jain AK, Prakah-Asante K,Prasad KV (2019) Actions Speak Louder Than (Pass) words: Passive Authentication of Smartphone Users via Deep Temporal Features. In: Proceedings of the 12th IAPR International Conference on Biometrics, Crete, Greece
23. Martinez-Diaz M, Fierrez J, Krish RP, Galbally J (2014) Mobile signature verification: feature robustness and performance comparison. IET Biometrics 3(4):267–277
24. O'Neal M, Balagani K, Phoha V, Rosenberg A, Serwadda A, Karim ME (2016) Context-aware active authentication using touch gestures, typing patterns and body movement. Louisiana Tech University, Ruston
25. Morales JF, Tolosana R, Ortega-Garcia J, Galbally J, Gomez-Barrero M, Anjos A (2016) Keystroke biometrics ongoing competition. IEEE Access 4:7736–7746
26. Perera P, Patel VM (2018) Efficient and low latency detection of intruders in mobile active authentication. IEEE Trans Inf Forensics Secur 13(6):1392–1405
27. Ernst R (2019) Mobile phone afterlife – why the second-hand market will be all the rage in 2019. In RCR Wireless News

Quickest Multiple User Active Authentication

Pramuditha Perera, Julian Fierrez, and Vishal M. Patel

Abstract In this chapter, we investigate how to detect intruders with low latency for Active Authentication (AA) systems with multiple-users. We extend The Quickest Change Detection (QCD) framework is extended to the multiple-user case and the Multiple-user Quickest Intruder Detection (MQID) algorithm is formulated. Furthermore, the algorithm is extended to the data-efficient scenario where intruder detection is carried out with fewer observation samples. The effectiveness of the method is evaluated on two publicly available AA datasets on the face modality.

1 Introduction

Balancing the trade-offs between security and usability is one of the major challenges in mobile security [4]. Longer passwords with a combination of digits, letters and special characters are known to be secure but they lack usability in the mobile applications. On the other hand, swipe patterns, face verification and fingerprint verification have emerged as popular mobile authentication methods owing to the ease of use they provide. However, security of these methods are challenged due to different types of attack mechanisms employed by intruders ranging from simple shoulder attacks to specifically engineered spoof attacks. In this context, Active Authentication (AA), where the mobile device user is continuously

P. Perera
Department of Electrical and Computer Engineering, Johns Hopkins University, Baltimore, MD, USA
e-mail: pperera3@jhu.edu

J. Fierrez
School of Engineering, Universidad Autonoma de Madrid, Madrid, Spain
e-mail: julian.fierrez@uam.es

V. M. Patel (✉)
Department of Electrical Engineering, Johns Hopkins University, Baltimore, MD, USA
e-mail: vpatel36@jhu.edu

T. Bourlai et al. (eds.), *Securing Social Identity in Mobile Platforms*, Advanced Sciences and Technologies for Security Applications,
https://doi.org/10.1007/978-3-030-39489-9_10

monitored and user's identity is continuously verified, has emerged as a promising solution [5, 20, 23].

Authors in [28] identified three characteristics that are vital to a practical AA system; accuracy, latency and efficiency. However, for AA to be deployed in the real-world, it needs to be equipped with another functionality – transferability. Mobile devices are not private devices that people use in isolation. In practice, it is common for mobile devices to be used interchangeably among several individuals. For example, these individuals could be the members of a family or a set of professionals operating in a team (such as physicians in a hospital). Therefore, it is important that the AA systems facilitate smooth transition between multiple enrolled individuals [26].

The presence of multiple enrolled subjects poses additional challenges to an AA system. Detecting intrusions with low latency in this scenario is even more challenging. With this new formulation, the device cannot simply declare an intrusion when there is a change in the device usage pattern. This is because two legitimate users operating on the phone could potentially have different behavior patterns. As a result, the systems is not only expected to identify intrusions, but also to provide smooth functioning when there is a transfer of legitimate users. For example, consider the scenario shown in Fig. 1. There are two legitimate users of the device in this scenario. The first user operates the mobile device between frames (a) and (c). At frame (d), the device is handed over to a second legitimate user. At this point, although there is a change in pattern in device usage, the AA system should not declare an intrusion. On the other hand, when an intruder starts using the device at frame (h), the device is expected to declare an intrusion.

In this chapter, we extend the work proposed in [28] and study the effectiveness of Quickest Change Detection (QCD) algorithm for multiple-user AA. Specifically, we study possible strategies that can be used to extend Mini-max QCD in AA to the case where multiple users are enrolled in the device. Furthermore, we study the effectiveness of data-efficient sampling for this case. In the experimental results section, we show that the QCD algorithm and it's data-efficient extension are effective even in the case of multiple-user AA.

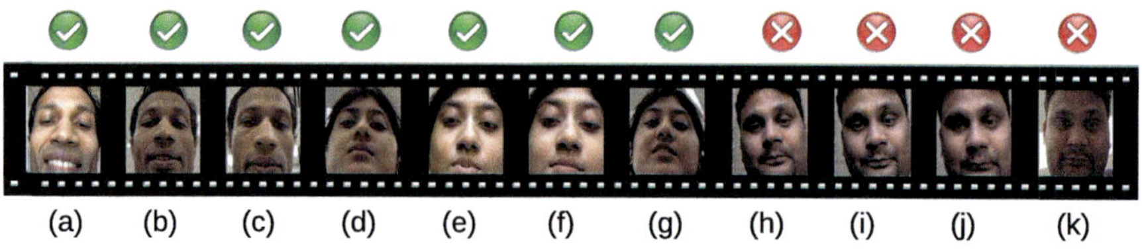

Fig. 1 Problem of quickest detection of intruders in multiple-user active authentication. In this example, there are two users enrolled in the mobile device. First user uses the device between frames (**a**) to (**c**). At frame (**d**), another legitimate user starts using the device. The second user uses the device between frames (**d**) to (**g**). At frame (**h**), an intruder starts using the device (In this work we assume that intruders do not attempt to hide their identity using a spoofing method). The goal of quickest intrusion detection is to detect the change with the lowest possible latency. However, intruder detector should not declare a false detection prior to frame (**h**)

2 Related Work

Initial works of AA predominantly focused on introducing new biometric modalities or increasing the performance of well-known modalities. Gait [15, 34], keystroke [9, 13], voice, swipe patterns [11, 30] and face images [6, 8, 17, 19, 21] are some of the commonly used modalities in mobile AA. In addition, micro movements of the user's touch gestures [3] and behavioral profiling based on stylometry, GPS location and web browsing patterns [12] have also been used for AA in the literature.

More recent works in AA focused on obtaining better authentication performance either by improving the performance in each individual modality or by fusing two or more biometric modalities. In [14], adaboost classifier and LBP features are used for face detection and face authentication in mobile devices. In [29], a facial attribute-based continuous face authentication was proposed for AA. A domain adaptive sparse dictionary-based AA system was proposed in [33], by projecting observations of different domains into a common subspace through an iterative procedure. McCool et al. [19] proposed to fuse face and voice data for obtaining more robust AA. In [6], face modality was fused with gyroscope, accelerometer, and magnetometer modalities for more robust authentication.

However, all of these methods focus on the single user authentication problem. Furthermore, the latency of decision making is not quantified in these works. In [26], the problem of single user AA was extended to the multiple user scenario. The authors proposed an SVM-based solution where the scores of each SVM output are fused using a new fusion rule. In speaker recognition, the need to have multiple user systems have been previously discussed [7, 18]. In [24] multiple user authentication is formulated as a conjunction between a classification task and a verification task. Based on the same principle, the authors of [27] introduced sparse representation-based intruder detection scheme for multiple-user AA. In [25], authors proposed to use the principles of QCD for AA. In [28], this formulation was extended with data-efficient QCD with the aim of producing highly accurate predictions with low latency while obtaining low number of sensor observations. In this work, we extend the algorithms presented in [28] and [25] to the multiple-user case and study it's effectiveness in face-based mobile AA.

3 Proposed Method

When a user or multiple users start using a mobile device, typically they are required to register with the device. This process is called enrollment of the user(s) to the mobile device. During enrollment, the device gathers sensor observations of the legitimate users and creates user-specific classifiers. Let m be the number of users enrolled in a given device. Technically, m could be any finite number greater or equal to one. However, in practice, it's not common for a mobile device to be shared between more than 5–7 individuals (i.e. normal family size).

For each user i, the device gathers enrollment data $Y_i = \{y_{i,1}, y_{i,2}, \ldots, y_{i,k}\}$. Then, the device develops a set of user specific classifiers c_i for each user which produces a classification score for each user. This classifier can be a simple template matching algorithm or a complex neural network. In our experiments, we consider a template matching algorithm due to the easiness in training the classifier. Our template matching classifier c_i generates a user specific score $s_i = c_i(y) = \min(\cos(y, Y_i))$ for a given input y where cos(.) is the Cosine angle between the two inputs.[1]

In addition, matched and non-match distributions with respect to the learned classifier are obtained and stored during the enrollment phase. Match distribution $f_{0,i}(.)$ of user i can be obtained by considering pairwise score values of Y_i with respect to c_i. On the other hand, a known set of negative samples can be used to obtain the non-matched scores $f_{1,i}(.)$ of user i. This process is illustrated in Fig. 2. In this work, we approximate the score distribution of intruders with the non-matched distribution. Therefore, we use the terms matched distribution and pre-change distribution interchangeably. Similarly, in the context of this paper, non-matched distribution and post-change distribution will also mean the same.

As the AA system receives observations $\{x_1, x_2, \ldots, x_n\}$, at time n, it produces a decision $d_n = f(C_1(x_1), \ldots, C_n(x_n)) \in \{0, 1\}$ based on the set of classifiers $C = \{c_1, \ldots, c_m\}$ where $f(.)$ is a mapping function. If $d_n = 1$, an intrusion is declared. Given this formulation, the goal of an AA system is to detect intrusions

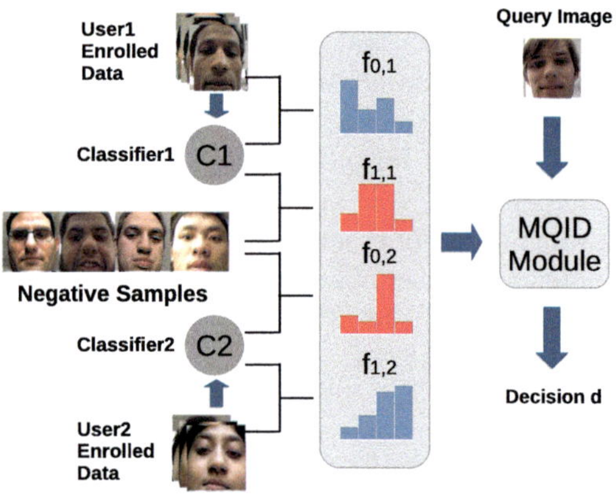

Fig. 2 Overview of the problem setup for the case of two enrolled users. For each enrolled user, i, the enrolled images are obtained during the enrollment phase. These images are used to train a user specific classifier c_i. A matched score distribution $f_{0,i}$ and a non-matched distribution $f_{1,i}$ is obtained for each user. A known set of negative users are used to obtain the latter. If more users are present the same structure will be cascaded. During inference, Multi-user Quickest Intruder Detection (MQID) module will produce a decision (d) by considering the obtained distributions and past decision scores

[1]Score s_i represents dissimilarity.

with the lowest possible latency when a new observation is received. If an intrusion occurs at time T, the following two properties are desired from the AA system.

- **Low detection delay.** The latency between an intrusion occurring and the system detecting the intrusion should be low. If the system detects an intrusion at time τ, detection delay is given by $(\tau - T)^+$ where $[(x)^+]$ denotes the positive part of x. For all users, this property is quantified using Average Detection Delay (ADD) defined as $ADD(\tau) = E[(\tau - T)^+]$. Here T denotes the real change point.
- **Low false detections.** In practice, detection delay alone cannot characterize the desired functionality of an AA system. It is also desired that the AA system does not produce false detections prior to the intrusion point. This phenomena can be quantified by considering Probability of False Detections (PFD) as $PFD(\tau) = P[\tau < T]$.

It is desired for an AA system to have low ADD and low PFD.

3.1 *Quickest Change Detection (QCD)*

Quickest Change Detection is a branch of statistical signal processing that thrives to detect the change point of statistical properties of a random process [1, 2, 31]. The objective of QCD is to produce algorithms that detect the change with a minimal delay (ADD) while adhering to false alarm rate constraints (PFD). Consider a collection of obtained match scores, $s_1, s_2, \cdots, s_n$, from the AA system. Assuming that the individual scores are mutually independent, QCD theory can be used to determine whether a change has occurred before time n or not. In the following subsections we present two main formulations of QCD.

QCD has been studied both in Baysian and a Mini-max frameworks in previous works. In the Baysian setting, it is assumed that the system has prior information about the distribution of intrusions. However, in the case of AA, probability of an intrusion happening cannot be modeled. Therefore, this assumptions does not hold. Therefore, for this work we only consider QCD in a non-Bayesian setting. MiniMax QCD formulation treats the change point τ as an unknown deterministic quantity [1, 2]. However, as mentioned before, it is assumed that pre-change distribution, f_0, and post-change distribution, f_1, are known.

Due to the absence of prior knowledge on the change point, a reasonable measure of PFD is the reciprocal of mean time to a false detection as follows

$$PFD(\tau) = \frac{1}{E_\infty[\tau]}.$$

Based on this definition of PFD, Lorden proposed a minimax formulation for QCD [2, 16]. Consider the set of stopping times D_α for a given constraint α such that $D_\alpha = \{\tau : PFD(\tau) \leqslant \alpha\}$. Adhering to this constraint, Lorden's formulation optimizes a cost function to solve the minimax QCD problem. In particular, the cost function is the supremum of the average delay conditioned on the worst possible realizations as follows

$$WADD(\tau) = \sup_{n \geqslant 1} \text{ess sup } E_n[(\tau - n)^+ | S^n].$$

Lorden's formulation tries to minimize the worst possible detection delay subjected to a constraint on PFD [16]. It was shown in [1], that the exact optimal solution for Lorden's formulation of QCD can be obtained using the CumSum algorithm [22].

3.1.1 CumSum Algorithm

Define the statistic $W(n)$ such that

$$W(n) = \max_{1 \leqslant k \leqslant n+1} \sum_{i=k}^{n} \log(L(s_i)),$$

and $W_0 = 0$, where $L(s_n) = f_1(s_n)/f_0(s_n)$ is the log likelihood ratio. It can be shown that the statistic $W(n)$ has the following recursive form

$$W_{n+1} = (W_n + \log(L(s_{n+1}))^+).$$

Time at which a change occurred (τ) is chosen such that $\tau_c = \inf\{n \geqslant 1 : W_n \geqslant b\}$, where b is a threshold. More details about the CumSum algorithm can be found in [1, 2, 22, 31].

3.2 *Efficient Quickest Change Detection*

Quickest Change Detection (QCD) is a branch of statistical signal processing that thrives to detect the change point of statistical properties of a random process [1, 2, 31]. The objective of QCD is to produce algorithms that detect the change with a minimal delay (ADD) while adhering to false alarm rate constraints (PFD). Consider a collection of obtained match scores, $s_1, s_2, \cdots, s_n$, from the AA system. Assuming that the individual scores are mutually independent, QCD theory can be used to determine whether a change has occurred before time n or not.

Consider a sequence of time instances $t = 1, 2, \cdots, i$ in which the device operates. At each time $i, i > 0$, a decision is made whether to take or skip an observation at time $i + 1$. Let M_i be the indicator random variable such that $M_i = 1$ if the score x_i is used for decision making, and $M_i = 0$ otherwise. Thus, M_{i+1} is a function of the information available at time i, i.e. $M_{i+1} = \phi_i(I_i)$, where ϕ_i is the control law at time i, and $I_i = [M_1, M_2, \cdots, M_i, s_1^{M_1}, s_2^{M_2}, \cdots, s_i^{M_i}]$ represents the information at time i. Here, $s_i^{M_i}$ represents s_i if $M_i = 1$, otherwise x_i is absent from the information vector I_i. Let S be the stopping time on the information sequence $\{I_i\}$. Then, average percentage of observations (APO) obtained prior to the change point can be quantified as $APO = E\left[\frac{1}{S}\sum_{n=1}^{S} M_n\right]$.

In a non-Bayesian setting, due to the absence of a priori distribution on the change point, a different quantity should be used to quantify the number of observations used for decision making. Work in [1, 2], proposes change Duty Cycle (CDC) as $CDC = \lim_n \sup \frac{1}{n} E_n\left[\sum_{k=1}^{n-1} M_k | \tau \geq n\right]$ for this purpose. It should be noted that both CDC and APO are similar quantities. With the definition of CDC, efficient QCD in a minimax setting can be formulated as the following optimization problem

$$\begin{aligned} \underset{\phi,\tau}{\text{minimize}} \quad & ADD(\phi, \tau) \\ \text{subject to} \quad & PFD(\phi, \tau) \leq \alpha, \quad CDC(\phi, \tau) \leq \beta. \end{aligned} \tag{1}$$

In [2], a two threshold algorithm called DE-CumSum algorithm, is presented as a solution to this optimization problem. For suitably selected thresholds chosen to meet constraints α and β, it is shown to obtain the optimal lower bound asymptotically as $\alpha \to 0$. The DE-CumSum algorithm is presented below.

Start with $W_0 = 0$ and let $\mu > 0$, $A > 0$ and $h \geq 0$. For $n \geq 0$ use the following control rule $M_{n+1} = 0$ if $W_n < 0$ otherwise 1 if $W_n \geq 0$. Statistic W_n is updated as follows

$$W_{n+1} = \begin{cases} \min(W_n + \mu, 0), & \text{if } M_{n+1} = 0 \\ \max(W_n + \log L(s_{n+1}), -h), & \text{if } M_{n+1} = 1, \end{cases}$$

where $L(s) = \frac{f_1(s)}{f_0(s)}$. A change is declared at time τ_W, when the statistic W_n passes the threshold A for the first time as $\tau_W = \inf\{n \geq 1 : W_n > A\}$.

3.3 Multi-user Quickest Intruder Detection (MQID)

Based on the discussion above, we introduce the Multiple-user Quickest Intruder Detection (MQID) algorithm. Whether an intrusion has occurred or not is determined using a score value. When the score value is above a pre-determined threshold, an intrusion is declared. At initialization, it is assumed that the user operating the device is a legitimate user; therefore the score is initialized with zero. The algorithm updates the score value when new observations are observed. During the update step, the algorithm considers matched and non-matched distributions of all users along with the current score value to produce the updated score. Pseudo code of the algorithm is shown in Algorithm 1.

The algorithm has three arguments. Argument *Efficient* determines whether to use data-efficient version of QCD or not. If data-efficient QCD is used then the parameter γ determines the floor threshold. Parameter D governs how fast the score is increased.

During training, enrolled images of each user along with the known negative dataset is used to construct matched and non-matched score distributions. In addition, enrolled images of the user are used to construct a classifier c_i. During inference, given an observation x, first classification scores from each classifier are obtained. Then, the likelihood of the obtained classifier score is evaluated using the likelihood ratio of each matched and non-matched distribution belonging to each user. The minimum likelihood ratio is considered as the statistic to update the current score of the system.

Updating the score based on the distribution is done as per the Algorithm considering the parameters as well as the magnitude of previous score value.

Algorithm 1: Algorithm to update the score based on the observations for the proposed method

```
input  : score, x_n, {f_{0,i}, f_{1,i}, c_i | ∀i}, γ, D, Efficient
output: score
L = min_i log( f_{1,i}(c_i(x_n)) / f_{0,i}(c_i(x_n)) )
if Efficient then
    if score < 0 then
        score = min(score + D, 0);
    else
        score ← max(score + L, −γ) ;
    end
else
    score ← score + L ;
end
Return (score);
```

4 Experimental Results

We test the proposed method on two publicly available Active Authentication datasets – UMDAA01 [8] and UMDAA02 [17] using the face modality. First, we explain the protocol used for evaluation. Then, we describe the performance metric used. Finally, we introduce the two datasets and present evaluation performance on these datasets (Fig. 3).

Fig. 3 Sample face images from the (**a**) UMDAA01 dataset and (**b**) UMDAA02 dataset used for evaluation. Samples from the same subject are shown in each column

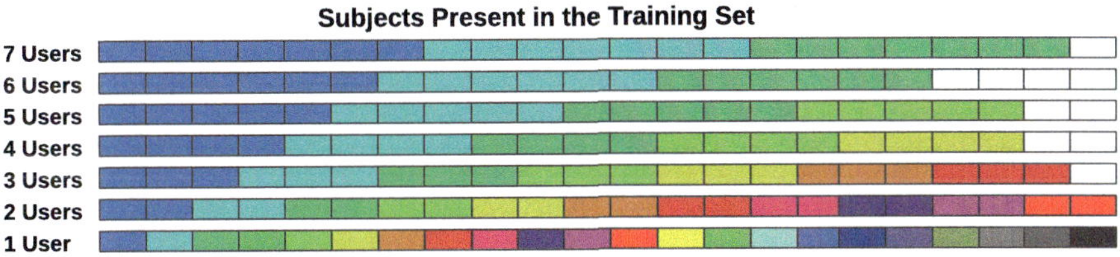

Fig. 4 Policy used to select the enrolled users for testing. The enrolled users considered together for a trial are denoted in the same color. For the case of 7 enrolled users, there are three trials. For the case of a single user, there are 22 trials

4.1 Protocol and Metrics

In both datasets, the first 22 users were used as possible enrolled users. Users 23–33 were used as the known negative samples. Remaining users were considered as intruders. From the enrolled users 10% of data were randomly chosen to represent the enrolled images. These image frames were removed from the test set. For each dataset, we varied the number of enrolled users from 1 to 7. If the number of enrolled users is m, we partitioned the first 22 users into disjoint groups of m and carried out floor$(22/m)$ trials. For example, in the case of 7 users, users 1–7 were considered to be enrolled in the first trial. For the second trial, users 8–14 were considered to be enrolled. Remaining users were not considered for testing. An illustration of the partitions obtained for several trials is shown in Fig. 4. In each trial, the intruder classes were considered one at a time and an intrusion was simulated.

In order to simulate an intrusion, the following process was followed. The entire video clips of the enrolled users were appended in the order of their index to form an augmented video for each trial. Then, each intruder from the intruder set was considered one at a time. Considered intruder's video clip was appended at the end of the augmented video clip to produce the test video clip. Shown in Fig. 1 is a summary of such a clip for the case of two enrolled users.

During training, we extracted the image frames from the video clip with a sampling rate of 1 image per 3 seconds. We used the Viola-Jones face detector to detect faces in the extracted image frame and performed local histogram normalization. The extracted image was resized to 224 × 224 and image features were extracted from the ResNet18 deep architecture which was pre-trained on the ImageNet dataset. For all cases, we considered the distance to the nearest neighbor as the user specific classifier c_i.

The performance of a quickest change detection scheme depends on ADD and PFD. Ideally, an AA system should be able to operate with low ADD and PFD. In order to evaluate performance of the system following [28], we used the ADD-PFD graph. We report ADD values required to obtain a PFD of 2% and 5% in Tables 1 and 2, respectively. These tables indicate the latency of detecting an intrusion in average while guaranteeing a false detection rate of 2% and 5%, respectively.

4.2 Methods

We evaluated the following methods using the protocol presented. For a fair comparison, in all cases except for in $P_n(FG17)$ we used the statistic $L = \min_i \log(\frac{f_{1,i}(c_i(x_n))}{f_{0,i}(c_i(x_n))})$ as the score value to perform intrusion detection.

Single score-based authentication (SSA) Present score value L alone is used to authenticate the user. If the score value is above a predetermined threshold, the user is authenticated otherwise treated as an intrusion.

Time decay fusion (Sui et al.) [32] In this method, two score samples fused by a linear function is used along with a decaying function to determine the authenticity of a user as, $s_n = wL_{n-1} + (1-w)L_n \times e^{\tau \delta t}$, where, w, τ are constants and δt is the time elapsed since the last observation.

Confidence functions (Crouse et al.) [6] A sequential detection score S_{login} is calculated by incorporating time delay since the last observation and a function of the present score x_n. The detection score is evaluated as, $S_{login,n} = S_{login,n-1} + f_{map}s_n + \int_{t_{prev}}^{t_{now}} f_{dec} dt$. Functions f_{map} and f_{dec} are empirical functions presented in [6].

Probability of Negativity (P_n(FG17)) [26] This method is proposed for multi-user AA. Matched and non-matched distributions of each user is used to produce an individual score. These uncertainty scores are then fused to produce the Probability of Negativity (P_n). For this baseline, we combined P_n score values sequentially using the method proposed in [32].

Multi-user Quickest Intruder Detection (MQID) The method proposed in this paper with the Min-Max formulation.

Data Deficient Multi-user Quickest Intruder Detection (DEMQID): The method proposed in this paper using the Min-Max formulation with data-efficient constraints. We selected parameter D by constraining the average number of observations to be 50% of all observations for the case of the single user. In our experiments we found this parameter to be 100.

4.3 Results

We carried experiments on the UMDAA01 and UMDAA02 datasets. The ADD-PFD curves are shown in Figs. 5 and 6 when the number of users are varied from 1 to 7. ADD values obtained for PFD of 2% and 5% are tabulated for UMDAA01 and UMDAA02 in Tables 1 and 2, respectively.

UMDAA01 Face Dataset The UMDAA-01 dataset [8] contains images captured using the front-facing camera of an iPhone 5S mobile device of 50 different individuals captured across three sessions with varying illumination conditions. Images of this dataset contain pose variations, occlusions, partial clippings as well as natural facial expressions as evident from the sample images shown in Fig. 3a. For our experiments we concatenated videos from all three sessions to form 50 classes.

In all considered cases MQID method has performed better than the other baseline methods when it was desired to achieve a PFD of 2%. It is also seen that $P_n(FG17)$, which is a method proposed for multi-user AA has also outperformed SSH method which uses log-likelihood ratio in all cases. Furthermore, data-efficient version of the algorithm, DEMQID, has performed on par with MQID, even performing better in certain cases. Average percentage of observations obtained in DEMQID for this dataset was 0.304.

However, it can be observed that when 5% of PFD is allowed, even other baseline methods perform reasonably well. For example, in majority of the cases SSH has performed on par with MQID. We also observe that DEMQID is slightly worse than MQID in this case. This suggests that for the employed deep feature, a PFD rate of 5% can be obtained even when the sequence of data are not considered. DEMQID takes more sparse samples when deciding the score value. As a result, when the score function is noisy, DEMQID is not affected by the noise as much as MQID. Even-though sparser sampling would result in some latency in detection, overall trade-off can be beneficial. This is why, DEMQID outperforms MQID when decision making is more challenging (as was the case when PFD of 2% was considered).

However, when the decision making becomes easier, DEMQID does not contribute towards improving the detection accuracy as score values are less noisy. This is why in the case of 5% of PFD, DEMQID performs worse than MQID.

UMDAA02 Face Dataset The UMDAA-02 Dataset [17] is an unconstrained multimodal dataset with 44 subjects where 18 sensor observations were recorded across a two month period using a Nexus 5 mobile device. Authors of [17] have made the face modality and the touch-data modality[10] publicly available. In our

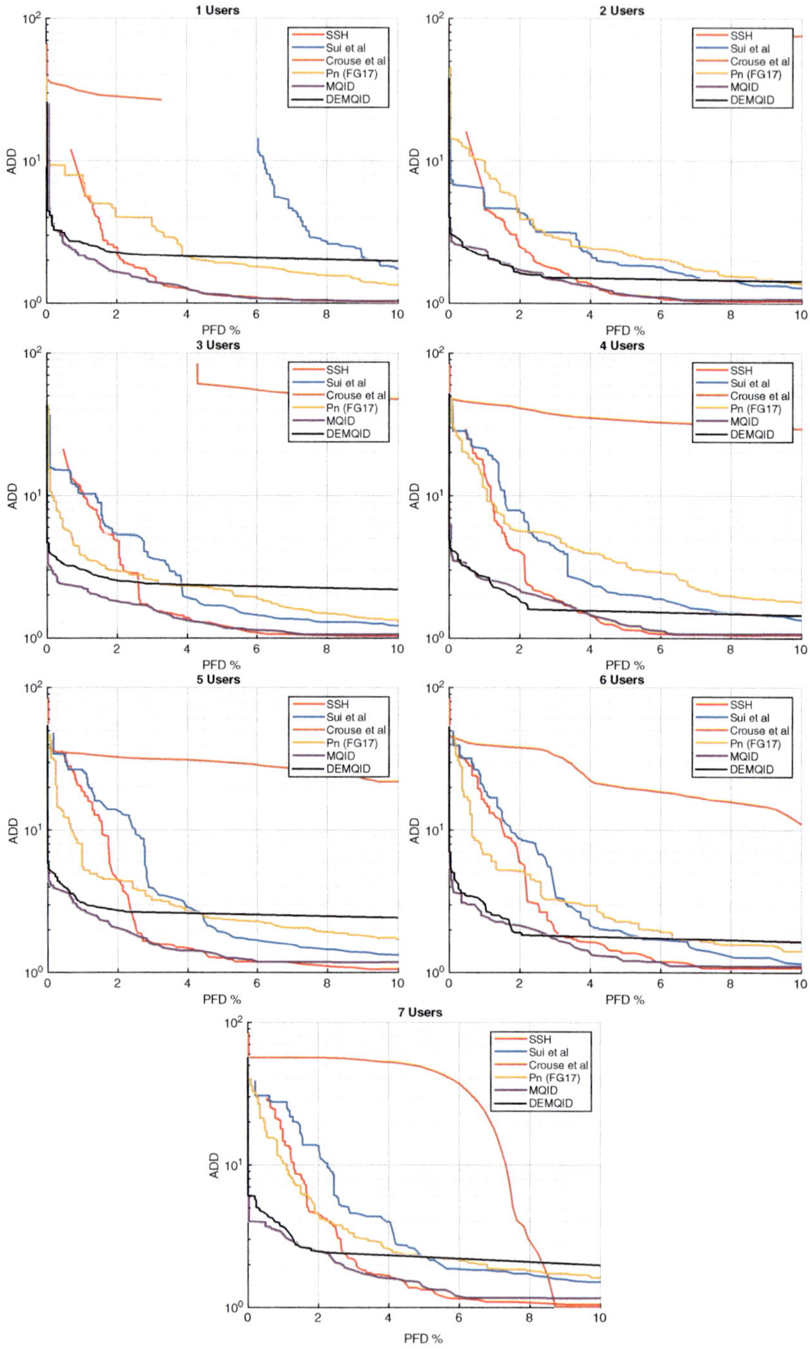

Fig. 5 The ADD-PFD curves corresponding to the UMDAA01 dataset when the number of users are varied from 1 to 7

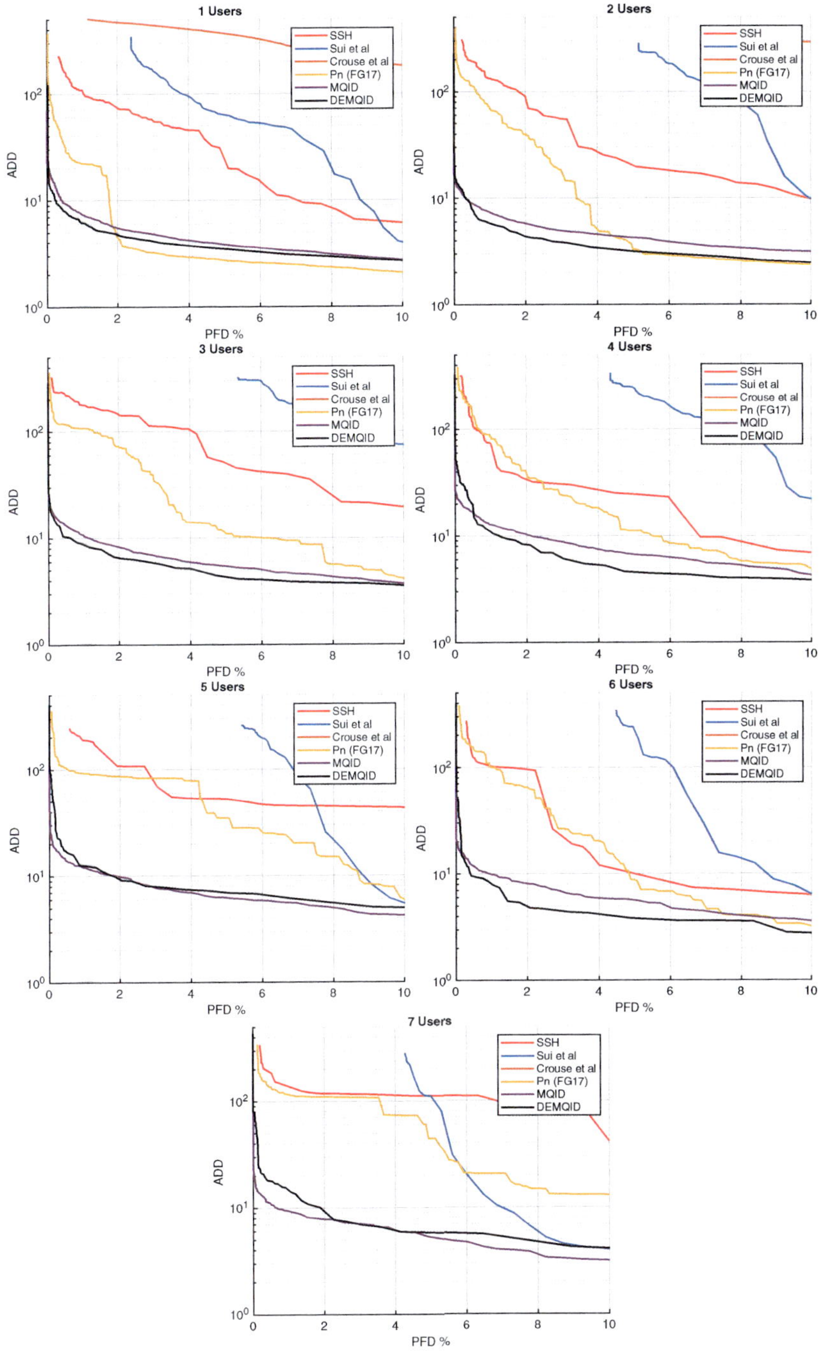

Fig. 6 The ADD-PFD curves corresponding to the UMDAA02 dataset when the number of users are varied from 1 to 7

Table 1 Tabulation of ADD for PFD of 2% and 5% when the users are varied from 1 to 7 on the UMDAA01 dataset. When a particular method failed to achieve prescribed PFD it is indicated by N/A

	5%							2%						
# of Users	1	2	3	4	5	6	7	1	2	3	4	5	6	7
SSH	1.14	1.14	1.17	1.18	1.25	1.35	1.31	2.28	2.49	4.73	3.98	4.41	5.84	4.49
Sui et al.	N/A	1.82	1.66	1.98	1.89	1.86	2.51	N/A	2.04	4.74	7.74	13.41	8.59	13.41
Crouse et al.	N/A	N/A	57.20	33.92	29.65	19.49	47.8	28.4	N/A	N/A	40.44	32.25	37.04	56.2
Pn (FG17)	2.10	2.20	2.16	3.29	2.35	2.23	2.51	3.96	3.84	2.89	5.48	4.41	5.06	4.51
MQID	1.14	1.14	1.17	1.20	1.25	1.19	1.31	1.63	1.65	1.79	2.08	2.02	2.12	2.49
DEMQID	2.14	1.52	2.37	1.59	2.51	1.86	2.51	2.28	1.64	2.51	1.84	2.72	1.92	2.49

Table 2 Tabulation of ADD for PFD of 2% and 5% when the users are varied from 1 to 7 on the UMDAA02 dataset. When a particular method failed to achieve prescribed PFD it is indicated by N/A

	5%							2%						
# of Users	1	2	3	4	5	6	7	1	2	3	4	5	6	7
SSH	19.7	19.6	51.82	24.33	52.64	10.01	110.2	72.92	90.71	150.0	36.23	107.6	93.6	118.1
Sui et al.	63.1	284.9	N/A	243.1	N/A	237.6	109.4	N/A	N/A	N/A	N/A	N/A	N/A	N/A
Crouse et al.	364.5	N/A	N/A	N/A	N/A	N/A	N/A	467.8	N/A	N/A	N/A	N/A	N/A	N/A
Pn (FG17)	2.71	3.31	10.91	11.22	34.64	7.06	44.55	4.30	39.56	72.52	37.44	86.3	64.0	116.0
MQID	3.83	4.28	5.42	6.67	6.11	5.61	5.30	5.58	5.77	8.14	10.38	9.10	8.03	7.78
DEMQID	3.47	3.17	4.13	4.618	6.93	3.85	5.82	4.32	4.39	6.38	8.34	9.12	4.79	10.15

work we only consider the face modality to perform tests. A sample set of images obtained from this dataset is shown in Fig. 3b. UMDAA02 is a more challenging dataset compared to UMDAA01 as apparent from the sample images shown in Fig. 3. In particular, we note the existence of a huge intra-class variations in this dataset in terms of poses, partial faces, illumination as well as appearances of the users.

As a result of having higher complexity, detecting intruders become more challenging in UMDAA02 compared to UMDAA01. However, due the challenging behavior of the dataset, the importance of the proposed method is magnified. In all ADD-PFD curves obtained for UMDAA02 in Fig. 6, it is evident that the proposed methods significantly outperform the baseline methods. Furthermore, DEQID has outperformed QID in most of the cases showing the significance of data efficient QCD.

In our evaluations we show that even when the number of users are increased, the performance of the proposed system does not drop drastically. For the UMDAA01 dataset, only 2.35 additional samples were required to maintain a probability of false detection of 2% when the users were increased from 1 o 7. In a more challenging UMDAA02 dataset, 4.33 more samples were required on average to maintain the same false detection rate.

5 Concluding Remarks

It has been previously showen that AA yields superior detection performance when the QCD algorithm is used [28]. In this chapter we study the problem of quickest change detection in a multiple-user AA scenario. We proposed MQID algorithm for multiple-user AA with low latency. Furthermore, we extended the initial formulation to a data efficient version by proposing DEMQID algorithm. We evaluated the performance of the proposed methods on the UMDAA01 and UMDAA02 datasets. Our experiments suggest that the proposed method is more effective compared to the baseline methods we considered. It was also shown that, the proposed method allows the number of enrolled users to be increased with a relatively smaller cost in terms of observations. Only 2.35 and 4.33 observations were required on average to maintain a false detection rate of 2% when the users were increased from 1 to 7 in the UMDAA01 and UMDAA02 datasets, respectively.

Acknowledgements This work was supported by the NSFgrant 1801435.

References

1. Banerjee T, Veeravalli VV (2014) Data-efficient quickest change detection. Sri Lankan J Appl Stat Special Issue: Modern Statistical Methodologies in the Cutting Edge of Science 183–208 Nov 2014

2. Banerjee T, Veeravalli VV (2013) Data-efficient quickest change detection in minimax settings. IEEE Trans Inf Theory 59:6917–6931
3. Bo C, Zhang L, Li X-Y, Huang Q, Wang Y (2013) Silentsense: silent user identification via touch and movement behavioral biometrics. In: Proceedings of the 19th annual international conference on mobile computing & networking, MobiCom '13. ACM, New York, pp 187–190
4. Crawford H, Renaud K (2014) Understanding user perceptions of transparent authentication on a mobile device. J Trust Manage 1(7):1–28
5. Crawford H, Renaud K, Storer T (2013) A framework for continuous, transparent mobile device authentication. Comput Secur 39:127–136
6. Crouse D, Han H, Chandra D, Barbello B, Jain AK (2015) Continuous authentication of mobile user: fusion of face image and inertial measurement unit data. In: International conference on biometrics
7. Dunn RB, Reynolds DA, Quatieri TF (2000) Continuous user authentication on mobile devices: recent progress and remaining challenges. IEEE Signal Process Mag 10(1–3):93–112
8. Fathy ME, Patel VM, Chellappa R (2015) Face-based active authentication on mobile devices. In: IEEE international conference on acoustics, speech and signal processing
9. Feng T, Zhao X, Carbunar B, Shi W (2013) Continuous mobile authentication using virtual key typing biometrics. In: Proceedings of the 2013 12th IEEE international conference on trust, security and privacy in computing and communications, TRUSTCOM '13, IEEE Computer Society, Washington, DC, pp 1547–1552
10. Fierrez J, Pozo A, Martinez-Diaz M, Galbally J, Morales A (2018) Benchmarking touchscreen biometrics for mobile authentication. IEEE Trans Inf Forensics Secur 13(11):2720–2733
11. Frank M, Biedert R, Ma E, Martinovic I, Song D (2013) Touchalytics: on the applicability of touchscreen input as a behavioral biometric for continuous authentication. IEEE Trans Inf Forensics Secur 8(1):136–148
12. Fridman L, Weber S, Greenstadt R, Kam M (2017) Active authentication on mobile devices via stylometry, Application usage, web browsing, and GPS location. IEEE Syst J 11(2):513–521
13. Gascon H, Uellenbeck S, Wolf C, Rieck K (2014) Continuous authentication on mobile devices by analysis of typing motion behavior. In: Katzenbeisser S, Lotz V, Weippl E (eds) Proceedings of GI conference Sicherheit, Bonn, Gesellschaft für Informatik e.V., pp 1–12
14. Hadid A, Heikkila JY, Silven O, Pietikainen M (2007) Face and eye detection for person authentication in mobile phones. In: ACM/IEEE international conference on distributed smart cameras, Sept 2007, pp 101–108
15. Juefei-Xu F, Bhagavatula C, Jaech A, Prasad U, Savvides M (2012) Gait-ID on the move: pace independent human identification using cell phone accelerometer dynamics. In: IEEE international conference on biometrics: theory, applications and systems, Sept 2012, pp 8–15
16. Lorden G (1971) Procedures for reacting to a change in distribution. Ann Math Stat 42(6):1897–1908
17. Mahbub U, Sakar S, Patel VM, Chellappa R (2016) Active authentication for smartphones: a challenge data set and benchmark results. In: IEEE international conference on biometrics: theory, applications and systems, Sept 2016
18. Martin AF, Przybocki MA (2001) Speaker recognition in a multi-speaker environment. In: INTERSPEECH
19. McCool C, Marcel S, Hadid A, Pietikainen M, Matejka P, Cernocky J, Poh N, Kittler J, Larcher A, Levy C, Matrouf D, Bonastre J-F, Tresadern P, Cootes T (2012) Bi-modal person recognition on a mobile phone: using mobile phone data. In: IEEE international conference on multimedia and expo workshops, July 2012, pp 635–640
20. Meng W, Wong DS, Furnell S, Zhou J (2015) Surveying the development of biometric user authentication on mobile phones. IEEE Commun Surv Tutorials 17(3):1268–1293
21. Narang N, Martin M, Metaxas D, Bourlai T (2017) Learning deep features for hierarchical classification of mobile phone face datasets in heterogeneous environments. In: 2017 12th IEEE international conference on automatic face and gesture recognition (FG 2017). IEEE Computer Society, Los Alamitos, pp 186–193

22. Page ES (1954) Continuous inspection schemes. Biometrika, 41(1/2):100–115
23. Patel VM, Chellappa R, Chandra D, Barbello B (2016) Continuous user authentication on mobile devices: recent progress and remaining challenges. IEEE Signal Process Mag 33(4): 49–61
24. Pelecanos J, Navrátil J, Ramaswamy GN (2008) Conversational biometrics: a probabilistic view. In: Ratha NK, Govindaraju V (eds) Advances in biometrics. Springer, London
25. Perera P, Patel VM (2016) Quickest intrusion detection in mobile active user authentication. In: International conference on biometrics theory, applications and systems
26. Perera P, Patel VM (2017) Towards multiple user active authentication in mobile devices. In: IEEE international conference on automatic face and gesture recognition
27. Perera P, Patel VM (2019) Face-based multiple user active authentication on mobile devices. IEEE Trans Inf Forensics Secur 14(5):1240–1250
28. Perera P, Patel VM (2018) Efficient and low latency detection of intruders in mobile active authentication. IEEE Trans Inf Forensics Secur 13(6):1392–1405
29. Samangouei P, Patel VM, Chellappa R (2015) Attribute-based continuous user authentication on mobile devices. In: IEEE international conference on biometrics: theory, applications and systems
30. Serwadda A, Phoha VV, Wang Z (2013) Which verifiers work? A benchmark evaluation of touch-based authentication algorithms. In: IEEE international conference on biometrics: theory, applications and systems, Sept 2013, pp 1–8
31. Veeravalli VV, Banerjee T (2012) Quickest change detection. ArXiv e-prints, Oct 2012
32. Zou X, Sui Y, Du EY, Li F (2012) Secure and privacy-preserving biometrics based active authentication. In: IEEE international conference on systems, man, and cybernetics (SMC), pp 1291–1296
33. Zhang H, Patel VM, Shekhar S, Chellappa R (2015) Domain adaptive sparse representation-based classification. In: IEEE international conference on automatic face and gesture recognition, vol 1, May 2015, pp 1–8
34. Zhong Y, Deng Y (2014) Sensor orientation invariant mobile gait biometrics. In: IEEE international joint conference on biometrics, Sept (2014), pp 1–8

Iris Recognition on Mobile: Real-Time Feature Extraction and Matching in the Wild

Gleb Odinokikh and Alexey Fartukov

Abstract Methods of biometric recognition are becoming an essential part of various mobile applications. Their usability is determined by the accuracy and the speed of recognition in a highly variable environment. Complex textural features make the human iris one of the most reliable biometric traits. The changing environment and limited computational power of mobile devices give rise to a need for robust and fast feature extraction techniques. A method for iris feature extraction and matching is here proposed. It uses deep and element-wise representations of the discriminative features in combination with characteristics describing the environment. The model outperforms state-of-the-art methods in terms of both accuracy and speed. It has also been tested on a specially collected dataset that contains two-second videos simulating the natural enrollment and verification attempts of the user of the device. The dataset was collected considering the changes in environment and possible behavior of the user. The testing was performed in two scenarios: image-to-image and also video-to-video. A method for iris fusion (both eyes) is also proposed in this paper. Several such methods are studied and compared.

Keywords Iris recognition · Mobile biometrics · Feature extraction · Matching · Multi-instance fusion

1 Introduction

Mobile devices, such as smartphones and tablets, have become an integral part of many people's lives. Nowadays, the transfer and processing of personal information and various financial transactions are carried out using mobile devices. They are personal, which means the presence of an authentication procedure.

G. Odinokikh (✉) · A. Fartukov (✉)
Samsung R&D Institute Russia, Moscow, Russia
e-mail: g.odinokikh@gmail.com; a.fartukov@samsung.com

T. Bourlai et al. (eds.), *Securing Social Identity in Mobile Platforms*, Advanced Sciences and Technologies for Security Applications,
https://doi.org/10.1007/978-3-030-39489-9_11

The methods of biometric authentication have been actively promoted to replace conventional schemes that use keys, personal identification numbers, etc. The growth of interest in biometric technologies is associated mainly with the strengthening of the security requirements of the system and its usability. A lot of attention has been paid to mobile biometrics in recent years [1–3].

The present paper is focused on the human iris as the most reliable biometric modality. The goal of iris recognition is to recognize a human's identity through the textural characteristics of the muscular patterns of the iris. A typical iris recognition system consists of the following stages: iris image acquisition, iris image segmentation, feature extraction, and pattern matching [4]. Iris image acquisition is usually performed using a high resolution camera which is either near-infrared (NIR) or visible-spectrum (VIS), under controlled environmental conditions [5]. The minimal requirements for iris image capturing are summarized in ISO/IEC 19794-6:2011 [6].

Since the mobile market is global, all the possible behavioral- and race-dependent features of the final users must be taken into account. For this reason, in particular, only the NIR spectrum is considered in this paper. The advantages of using NIR images have been well explained in the literature [6–9]. It should be noted that the development and implementation of an iris capturing camera for mobile devices is outside the scope of the present paper. Most of the issues related to the iris capturing device are well summarized in [5, 7].

In the case of a mobile device, it is not always possible to satisfy all the mentioned requirements imposed on the camera. There are two main reasons for this: the camera module should be small enough and not expensive to manufacture. The costs of production play a significant role in the case of a mass market. Another challenge is that capturing the iris image is performed under uncontrolled environmental conditions. These factors greatly affect the quality of the iris image.

This paper describes a method of iris feature extraction and matching that is capable of working in real-time on a mobile device equipped with an NIR camera.

The rest of this paper is organized as follows: the key issues of iris recognition on a mobile device are explained in Sect. 2; Sect. 3 surveys related work; the proposed approach is presented in Sect. 4; and experimental results and conclusions are presented in Sects. 5 and 6 respectively.

2 Problem Statement

A mobile biometric sensor should be able to handle the data under constantly changing environmental conditions and consider user inherent features. In biometric systems that use an image as input data, the factors of the environment are becoming more important. One of them is the ambient illumination, which varies over a range from 10^{-4} at night to 10^5 Lux under direct sunlight. Another is the randomness of the locations of the light sources, along with their unique characteristics, which creates a random distribution of the illuminance in the iris area. These factors lead to a deformation of the iris structure caused by a change in the pupil size,

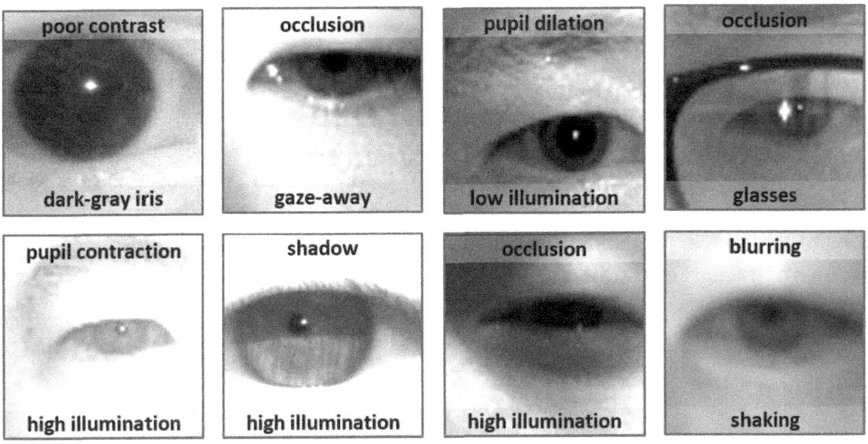

Fig. 1 Examples of iris images captured with a mobile device

making users squint, and degrading the overall image quality. Several examples of iris images are given in Fig. 1. The influence of the environment is well described in the literature [10–13]. Other factors inherent to the user also affect the quality of the output, such as the use of glasses or contact lenses, the existence of a hand tremor, or the mere act of walking, thereby introducing a shaking of the device, the variation in distance to the iris causing the iris to move out of the camera's depth of field, and occlusion of the iris area by eyelids and eyelashes if the user's eye is not opened enough [10]. All these and many other factors affect the quality of the input biometric data thus influencing the accuracy of the recognition [14, 15].

Mobile applications should be simple and convenient in use. In the case of a biometric system on a mobile device, being convenient means providing an easy user interaction and a high recognition speed, which is determined by the computational complexity. Mobile secure systems that process any personal data are even more limited in computational resources. Not many researchers have attached importance to this. These systems typically represent a system-on-a-chip (SoC) completely abstracted from external resources, keeping all the processing inside itself. Such systems were initially developed to carry out simple operations with simple data (PINs, passwords etc.) and did not require huge resources. Neither were they ready for the complex processing of biometric information, but they have continued to be improved. The latter system' restrictions usually meant even more reduced CPU frequency and limited RAM.

All these problems greatly complicate iris feature extraction, making most of the existing methods unreliable, and promising techniques such as deep neural networks (DNN) inoperable.

There are several commercial mobile iris recognition solutions known to date. The first smartphones enabled with the technology were introduced by Fujitsu [16] and Microsoft [17] in 2015. All Samsung flagship devices were equipped with iris recognition technology during 2016–2018 [18]. Some B2B and B2G applications of the technology are also known in the market, such as Samsung Tab Iris [19] and IrisGuard EyePay Phone [20]. The scope of the applications of this technology is growing and has brought about a demand for its further improvement.

The present paper is focused on the feature extraction and matching parts of the iris recognition pipeline. Feature extraction, in this case, means a numeric representation of the unique iris features extracted from the preliminarily determined iris area of the image. Matching means calculating a measure of dissimilarity between the two extracted feature vectors.

3 Related Work

Recent achievements in the field of deep learning have allowed a significant leap in the reliability and quality of the research in the field of biometrics and, in particular, in iris recognition. One of the first attempts to explore the capabilities of DNNs was a feasibility analysis of DNN embeddings trained on ImageNet for classification, with the PCA+SVM applied over the VGG embeddings [21] by Minae et al. Furthermore, Gangwar et al. [22] introduced their DeepIrisNet as a model combining all successful deep learning techniques known at the time. They thoroughly investigated the obtained features and produced a strong baseline as a robust foundation for future research. A year later, similar work based on these embeddings was introduced by Tang et al. [23]. At the same time, Proenca et al. [24] presented IRINA. The idea was to use a DNN to find corresponding patches from the examined images, use MRF to perform precise deformable registration, and a SVM to classify genuine and impostor data. They achieved unprecedented robustness to pupil/iris variations and segmentation errors, but the accuracy of the solution was traded off against performance. The proposed design significantly limited the applicability of the method for mobile applications. Another approach with two fully-convolutional networks with a modified triplet loss function has been proposed recently [25]. One of the networks is used for iris template extraction whereas the second produces the accompanying mask. Fuzzy image enhancement combined with simple linear iterative clustering and an SOM neural network were proposed in [26]. Although this method was designed for iris recognition on a mobile device, real-time performance was not achieved. Another recent paper [13] meant to be suitable for the case of a mobile device proposed a two-headed (iris and periocular) CNN with a fusion of the embeddings. Thus, there is no fully optimal solution for iris feature extraction and matching in the published papers.

4 Iris Feature Extraction and Matching

The proposed method represents a CNN designed to use the advantages of the normalized iris image as an invariant, both low and high level discriminative feature representations, and information about the environment. It contains iris feature extraction and matching parts trained together.

4.1 Recognition Pipeline

A common iris recognition pipeline consists of several stages separated by intermediate quality checks. The feature extraction part is preceded by the segmentation stage and followed by the matching. All the input data for the feature extraction (normalized iris and mask images) were obtained automatically by an algorithm developed in our lab. The basic structure of the algorithm was taken from [10] with two modifications: (i) the scheme that contains a special quality buffer was replaced with a straightforward structure as depicted in Fig. 2; (ii) the feature extraction and matching parts were also replaced with the new ones. All the other parts of the algorithm and quality checks were used with no modifications.

4.2 Low-Level Feature Representation

It is known from the previous literature [27, 28] that shallow layers in CNNs are responsible for the extraction of low-level textural information while high-level

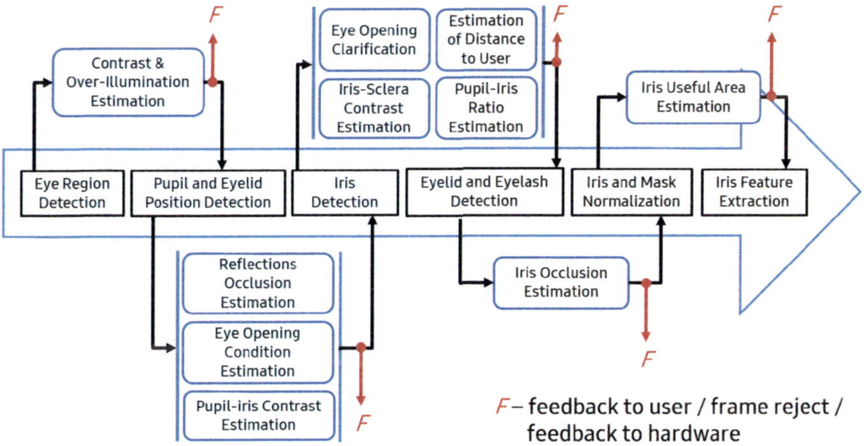

Fig. 2 Quality assessment scheme

representation is achieved with depth. Methods of iris feature extraction based on local texture characteristics, which are calculated by spatially and spectrally local transformations [9, 29] are basically attempts to use low-level description of the texture. These methods have proven their reliability for scenarios with an almost unchanging environment, but are highly sensitive to environmental variations.

A normalized image of the iris allows textural element-wise features to be useful in the case of narrow changes of environmental conditions. They remain well aligned with each other in such a case. For this reason, iris recognition is a good example of a task for which the profitability of using low-level feature representations could be investigated in the context of CNN-based methods and a wide range of environmental changes.

The influence of shallow features in the context of CNNs on recognition performance is studied in this paper. A classic approach [9] using a Hamming-distance based dissimilarity score has been taken as the basis. The vector FV_{sh} of elements x_i is used as a representation of low-level discriminative features:

$$x_i = \frac{\sum |FM_{1,i}^{Sq} - FM_{2,i}^{Sq}| \times M_c}{\sum M_c}, \tag{1}$$

where $FM_{k,i}^{Sq}$ is the ith feature map of the kth iris after normalization to zero mean and unit variance, binarized by sign; M_c is a binary mask representing noise; and M_c is a combination of M_1 and M_2.

The shallow feature extraction block is depicted in Fig. 3 and the structure of the convolution block #1 is presented in Table 1. Depth-wise separable convolution block structures, first proposed in [30] as memory and computationally efficient, were picked as the basic structural elements for the entire network. Feature maps $FM_{1,i}^{Sq}$ and $FM_{2,i}^{Sq}$ in (1) are obtained after the first convolution layer (Table 1).

After 100 epochs of training, the distributions of the elements of FV_{sh} for genuine and impostor comparisons from the validation set appear as in Fig. 4. Although the filters vary considerably, the distributions look very similar. The shape of the distributions for both classes resemble a Gaussian, therefore d' and EER values were chosen for the evaluation of their separation degree. How the values for each filter were changed during training is presented in Fig. 5. The results presented in Table 3 show that the model using FV_{sh} as an additive factor obtains slightly better results for the baseline model with 3×3 kernels on the first layer. It is also shown that for the larger kernels, the difference in performance becomes more significant (Table 3).

4.3 Deep Feature Representation

Deep (high-level) feature representation is obtained with convolution block #2. The feature maps $FM_{1,i}^{Sq}$ and $FM_{2,i}^{Sq}$ are concatenated after block #1 by channels

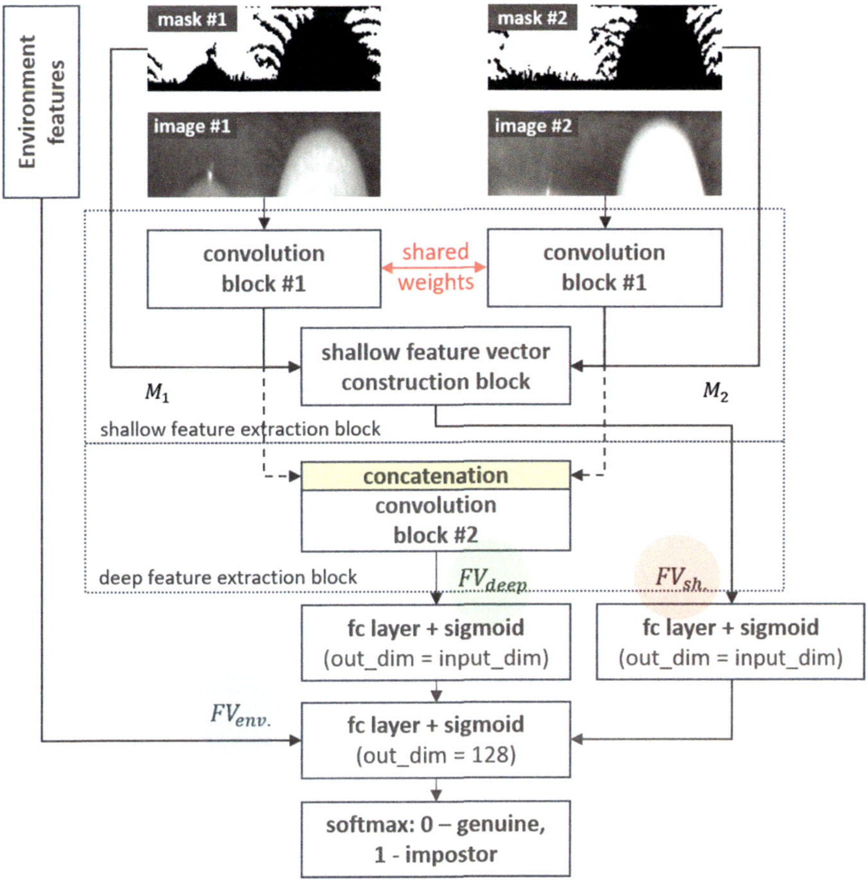

Fig. 3 Proposed model scheme

Table 1 Structure of the convolution blocks

Layer	Input shape	Convolution block
Conv. 3x3 ($s' = 1, act. = tanh$)	$1 \times 49 \times 161$	#1
Depthwise Sep. Conv. Block ($s' = 2$)	$8 \times 47 \times 159$	
Depthwise Sep. Conv. Block ($s' = 2$)	$32 \times 23 \times 79$	#2
Depthwise Sep. Conv. Block ($s' = 2$)	$32 \times 11 \times 39$	
Depthwise Sep. Conv. Block ($s' = 1$)	$32 \times 5 \times 19$	
FC layer + BatchNorm + ReLU	1×1632	

and passed through it (Fig. 3). The meaning of the concatenation at this stage is in the invariance property of the normalized iris image. Experiments showed the advantages of this approach in comparison with standard techniques [31] where the feature vectors had highly decreased dimensionality. However, the large sizes of the

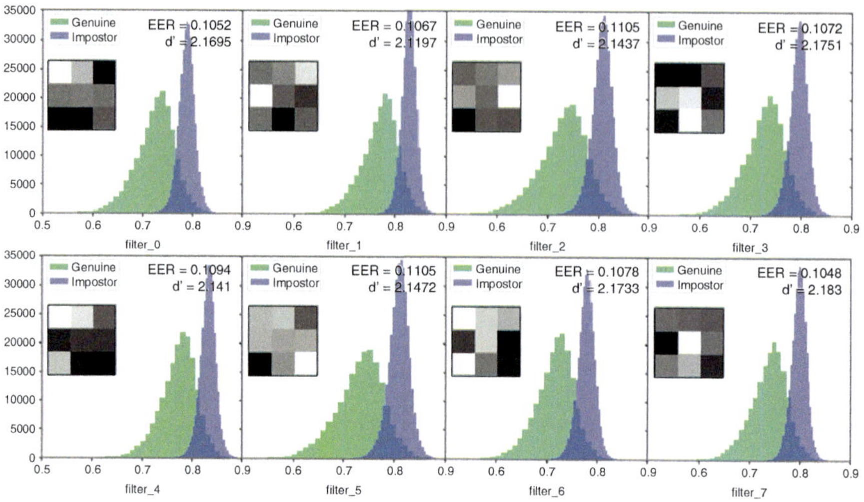

Fig. 4 Distributions of elements of FV_{sh} after 100 epochs

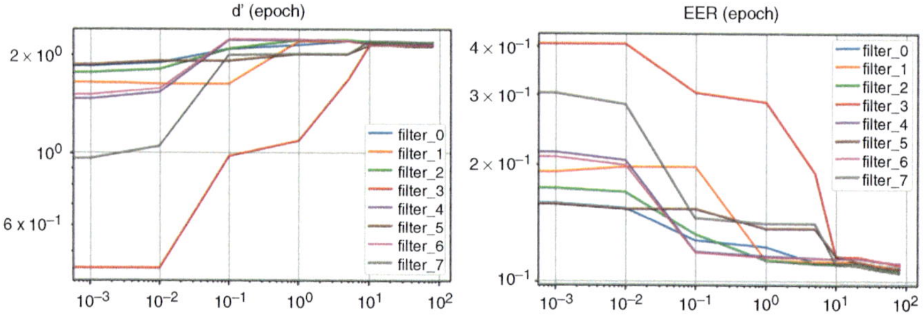

Fig. 5 The dependency of d' (left) and EER (right) for each filter on the number of epochs

vectors and the complexity of the matching procedure are among the drawbacks of this approach. The structure of the block is presented in Table 1. The output vector $FV_{deep} \in R^{128}$ reflects a high-level representation of the discriminative features and is assumed to handle complex non-linear distortions of the iris texture.

4.4 Matching Score Calculation

The analysis of outliers along with the nature of the distributions of the elements of FV_{sh} gave rise to the idea of using a variational inference technique for regularization. What this means is that some vectors are being represented as n-dimensional random variables with a certain shape distribution. In the present paper, the representations of both FV_{sh} and FV_{deep} vectors are described as having multivariate normal distributions $FV'_{sh} \sim N(\mu_{sh}, \Sigma_{sh})$ and $FV'_{deep} \sim N(\mu_{deep}, \Sigma_{deep})$ respectively, where μ is the vector of mean values and Σ is the covariance matrix. Variational inference is performed with the so called re-parametrization trick described in [32]. Sampling from the distributions is performed only for training, while only the values of μ are used for inference. A sigmoid activation function is then applied to the result. The same procedure is further performed for the concatenated vectors FV'_{sh}, FV'_{deep} and FV_{env}. Here, FV_{env} reflects environment conditions and contains information about iris area and pupil dilation: $FV_{env} = \{\Delta NPR, AOI\}$, and the area of intersection $AOI = \Sigma M_c / M_c^h \times M_c^w$ with ΔNPR given by

$$\Delta NPR = \left| \frac{R_1^p}{R_1^i} - \frac{R_2^p}{R_2^i} \right| \tag{2}$$

where R^p and R^i are the radii of the pupil and the iris, respectively. The output vector $FV'_d \in R^{128}$ is an input for the last fully-connected layer with two nodes describing the classes. A *SoftMax* classifier is applied to the values from the nodes for probability (matching score) estimation.

According to the obtained results (Table 3), the application of variational inference (VI) improved the recognition performance (VI = No means the replacement of the VI structure with simple fully-connected layers of the same dimensionality and activations), but it is also worth mentioning that it becomes less reasonable with an increasing amount of training data.

4.5 Weighted Loss

A specially designed loss function is another proposed feature. Sometimes two images of the same iris are very different from each other. This can happen for various reasons: the different parts of the iris can be occluded by some noise, one of the images can be badly distorted due to segmentation errors, etc. Thus, it is almost impossible to attribute them to the same class and for this reason a certain part of all genuine comparisons in the training data obstruct the convergence of the model. So, it is reasonable to consider or even completely ignore these comparisons when training. The following algorithm is proposed: (i) calculate the loss function (e.g., cross-entropy) for each comparison in the batch; (ii) apply $weights = \{w_0..w_K\}$ to the top k highest values among the genuine matches; (iii) output the overall sum. In this paper, the values were set to: $weights = 0$ and $k = 10\%$. This approach provided better convergence and achieved a better recognition performance.

4.6 Multi-instance Iris Fusion

The input images may contain both eyes, as depicted in Fig. 6. In this case both irises can be used at the same time [33], which is the obvious way to increase the reliability and convenience of the recognition. It has been observed that at least 40% of the iris area should be visible to achieve the given accuracy level. In other words, the user should open the eyes wider during one-eye recognition, which is not always convenient. Often the iris is significantly occluded by the eyelids, eyelashes, highlights, etc. This happens mainly because of the complex environment, in which the user cannot open the eyes wide enough (bright illumination, windy weather, etc.). It makes the application of the iris multi-instance approach reasonable.

An ideal scenario for matching is when both compared irises are well aligned to each other spatially and the conditions of the capturing are the same in both cases [15, 34]. This is impossible to satisfy in practice, especially in the mobile case. But it is reasonable to use information about the initial relative position of the compared irises before the normalization. A method that performs the fusion of the two irises and uses the relative spatial information and several factors that describe the environment is also considered as an important path of the presented research.

The final dissimilarity score is calculated as a logistic function of the form

$$score = \frac{1}{1 + e^{-\sum w_i \cdot M_i}} \tag{3}$$

where $M \in R^7$ is the set of the following measures:

$$M = \left\{\Delta d_0, d_{avg}, AOI_{min}, AOI_{max}, \Delta ND_{min}, \Delta ND_{max} \Delta PIR_{avg}\right\} \tag{4}$$

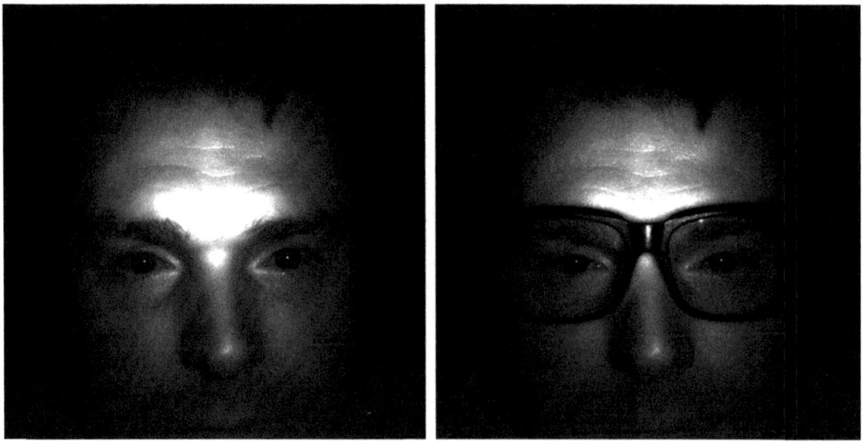

Fig. 6 Examples of the images captured with the mobile device equipped with an NIR camera

where

Δd_0 is the normalized score difference for two pairs of irises,

$$\Delta d_0 = \frac{\left| d_0^{left} - d_0^{right} \right|}{d_0^{left} + d_0^{right}} \tag{5}$$

d_{avg} is the average score for the pair,

$$d_{avg} = 0.5 \cdot (d_0^{left} + d_0^{right}) \tag{6}$$

AOI_{min}, AOI_{max} are the minimum and maximum values of the area of intersection between the two binary noise masks M_{prb} and M_{enr} in each pair,

$$AOI = \Sigma M_c / (M_c^h \times M_c^w),\ M_c = M_{prb} \times M_{enr} \tag{7}$$

ΔND_{min}, ΔND_{max} are the minimum and maximum values of the normalized distance ΔND between the centers of the pupil and the iris,

$$\Delta ND = \sqrt{(NDX_{prb} - NDX_{enr})^2 + (NDY_{prb} - NDY_{enr})^2} \tag{8}$$

$$NDX = \frac{x_P - x_I}{R_I}, NDY = \frac{y_P - y_I}{R_I}, \tag{9}$$

where x_P and y_P are the coordinates of the center of the pupil and R_P is its radius, while x_I and y_I, are the coordinates of the center of the iris and R_I is its radius, as depicted in Fig. 7.

The measure ΔPIR_{avg} reflects the difference in pupil dilation during the enrollment and probe using the value of $PIR = R_P / R_I$:

$$\Delta PIR_{avg} = 0.5 \cdot \left(\left| PIR_{enr}^{left} - PIR_{prb}^{left} \right| + \left| PIR_{enr}^{right} - PIR_{prb}^{right} \right| \right) \tag{10}$$

where R_P and R_I are the radii of the pupil and the iris, respectively.

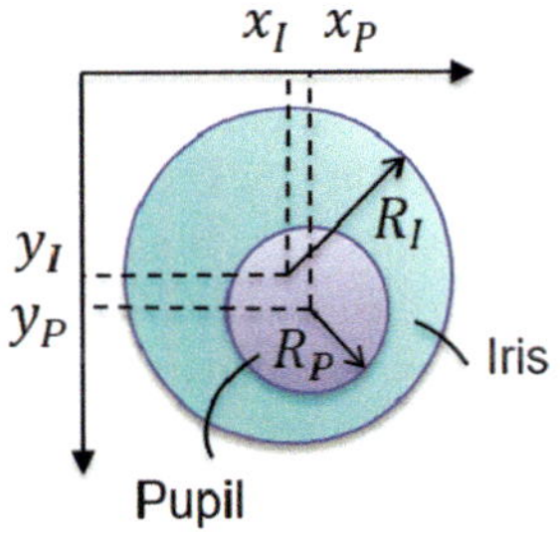

Fig. 7 Parameters of the pupil and iris used for the iris fusion

The weight coefficients for the logistic function were obtained after the training of the classifier on genuine and impostor matches on a small subset of the data. In case only one out of two feature vectors is extracted, all the pairs of values used in the weighted sum are assumed to be equal.

The proposed method helped to increase the recognition accuracy. It is also allowed to decrease the threshold for the visible iris area from 40% to 29% during verification/identification without any loss in the accuracy and performance, which means a decreased overall FRR as a result.

A comparison of the proposed method with well-known consensus and minimum rules was carried out. According to the consensus rule, a matching is considered as successful if both d_0^{left} and d_0^{Right} are less than the decision threshold. In the minimum rule, what is required is that the minimum of the two values $\min(d_0^{left}, d_0^{right})$ should be less than the threshold. The testing results are presented in Table 7.

5 Experimental Results

The main objectives of the biometric system performance evaluation include assessing the progress in improving the accuracy during the development of the algorithms and providing an objective reflection of the performance when the system is in operation [35]. To meet these goals, two types of evaluation were conducted: (i) an image-to-image evaluation of the proposed feature extraction and matching method with state-of-the-art methods on several datasets, including publicly available ones; (ii) a video-to-video evaluation to simulate real-world usage of the whole iris recognition solution and test the proposed multi-instance iris fusion approach.

5.1 Image-to-Image Evaluation

Three different datasets were used for the comparison. The following methods were selected as state-of-the-art: (1) FCN+ETL proposed by Zhao and Kumar in [25], which is one of the most cutting edge solutions, with the highest recognition performance; (2) DeepIrisNet [22], representing a classic deep neural net approach as one of the earliest applications of deep learning in the field of iris recognition. A lightweight CNN recently proposed in [13] could also be used for comparison since the results were obtained on the same dataset. Refer to the original paper [13] for the results on the CASIA-Iris-M1-S3 dataset [36].

Many methods were excluded from consideration due to their computational complexity and therefore unsuitability for mobile applications.

Table 2 Datasets details

Dataset	Images	Irises	Outdoor	Subjects
CMS2	7723	398	0	Asian
CMS3	8167	720	0	Asian
IM	22966	750	4933	Asian & Caucasian

5.1.1 Dataset Description

The following datasets were used for training and evaluation: CASIA-Iris-M1-S2 (CMS2) [36], CASIA-Iris-M1-S3 (CMS3) [36], and Iris-Mobile (IM). The collection of the last one was performed privately using a mobile device with an embedded NIR camera to simulate real authentication scenarios of the user of a mobile device. The images were captured under a wide range of changes in illumination, both indoors and outdoors, with and without glasses (Table 2). Examples of images are presented in Fig. 1. Images from all the datasets were marked automatically by an algorithm developed in our lab. Examples of iris and mask images are presented in Fig. 3. Each dataset was initially divided into training, validation, and testing subsets in proportions of 70/10/20 (%) respectively. This was so that there were would be no images of the same iris in two different subsets.

5.1.2 Training

The number of genuine comparisons N_G was much smaller than the number of impostor comparisons. Therefore all genuine comparisons were used for training and the number of impostor comparisons was fixed as $N_I = 10N_G$. The model that showed the lowest EER on the validation set was selected for evaluation on the testing dataset. All the models were trained for 150 epochs using the Adam optimizer. The training of the proposed model was performed so that one epoch was equivalent to one iteration over all the genuine comparisons whereas the impostors are always randomly selected from the entire set for each batch. The proportion of genuine and impostor comparisons in a batch was set to $N_I^b = 10N_G^b$ and $AOI \geq 0.2$ for all the image pairs.

5.1.3 Performance Evaluation

The recognition accuracy results are presented in Table 4 and Fig. 8. The proposed feature extraction and matching method outperforms the chosen state-of-the-art ones on all the datasets. Since the number of comparisons for the CMS2 and CMS3 testing sets did not exceed 10 million after the division into subsets, it was impossible to estimate the FNMR at FMR = 10^{-7}. Another experiment was used to estimate the performance of the proposed model on those datasets without training on them. The model trained on IM was evaluated on the entire CMS2 and CMS3

Table 3 Recognition performance results for several model modifications on IM dataset

Conv1	VI	FV_{sh}	EER	FNMR	d'
$8 \times 3 \times 3$	Yes	Yes	0.0116	0.1925	4.3155
$8 \times 3 \times 3$	No	Yes	0.0120	0.2027	4.2048
$8 \times 3 \times 3$	Yes	No	0.0125	0.2085	4.1253
$8 \times 9 \times 9$	Yes	Yes	0.0134	0.1566	4.3034
$8 \times 9 \times 9$	Yes	No	0.0172	0.1694	3.9850

Table 4 Recognition performance evaluation results

	EER				
Method	CMS2	CMS3	IM	Testing	FPS
DeepIrisNet [22]	0.0709	0.1199	0.1371	WithinDB	11
FCN+ETL [25]	0.0093	0.0301	0.0607	WithinDB	12
Proposed	0.0014	0.0190	0.0116	WithinDB	250
	0.0003	0.0086	–	CrossDB	

datasets in order to obtain FNMR at FMR = 10^{-7} (CrossDB). The results presented in Table 4 and Fig. 8 demonstrate the high generalization ability of the model. However, it is fair to note that the IM dataset contains much more data than the other two.

A mobile device equipped with Qualcomm Snapdragon 835 CPU was used for estimating the overall execution time for these iris feature extraction and matching methods. It should be noted that a single core of CPU was used. The results are summarized in Table 4.

5.2 *Video-to-Video Evaluation*

In fact, both the registration and verification procedures involve the processing of not one, but a sequence of images. The video format gives more information about the possible behavior of the user and the environment. Unfortunately, there are no such publicly available iris datasets. So, in order to test the recognition performance on data that would be close to real-world scenarios, an additional dataset was collected privately. It is a set of two-second video sequences, each of which is a real enrollment/verification attempt.

5.2.1 Dataset Description

The dataset was collected using a mobile device with an embedded NIR camera. It contains videos captured in different environment: (i) indoors (IN) and outdoors (OT); (ii) with and without glasses; (iii) at different distances. The conditions of illumination during the capturing were set as: (i) three levels for the indoor samples

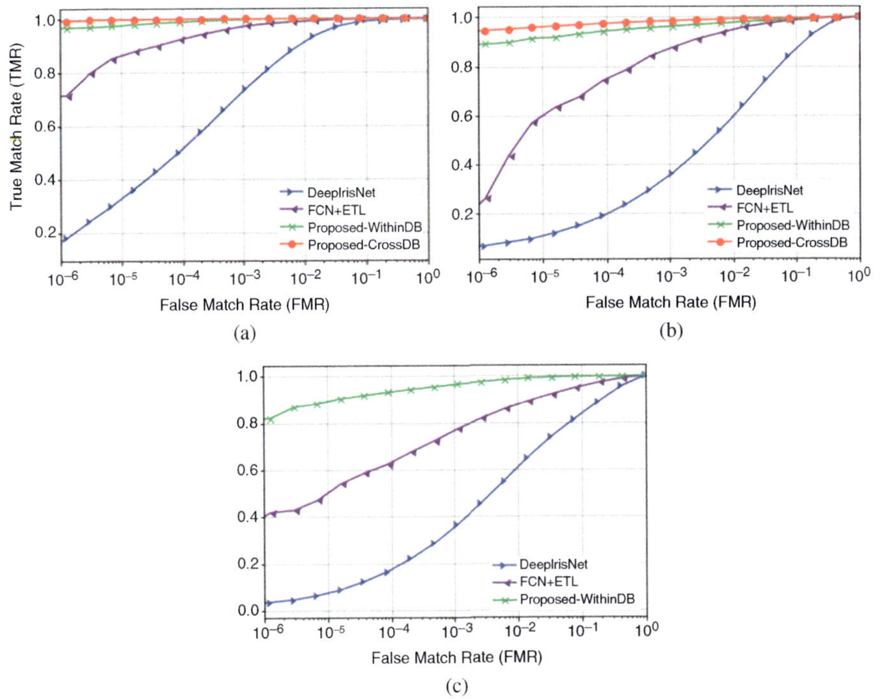

Fig. 8 ROC curves obtained for comparison with state-of-the-art methods on different datasets: (**a**) CASIA-Iris-M1-S2 (CMS2), (**b**) CASIA-Iris-M1-S3 (CMS3) and (**c**) Iris Mobile (IM)

Table 5 Dataset specification

Dataset	Non-glasses	Glasses
Users in dataset	476	224
Max comparisons	22 075 902	10 605 643
Race	Asian & Caucasian	
Eyes on video	Two	
Videos per user	10 ± 2	
Video length	30 frames	
Capturing distance	25–40 cm	
Camera resolution	1920×1920	

(0–30, 30–300 and 300–1000 Lux); (ii) a random value in the range 1–100 K Lux (data was collected on a sunny day); Different arrangements of the device relative to the sun were also considered during the capturing. A detailed description of the dataset is presented in Table 5. The Iris Mobile (IM) dataset used for the image-to-image evaluation was randomly sampled from, as well. Examples of the pictures from the videos are depicted in Fig. 6.

5.2.2 Testing Procedure

All the video sequences were used for simulating both the enrollment and verification transactions (attempts) in the non-glasses (NG) case. The sequences captured for users wearing glasses (G) were used for simulation of the verification attempts only. Each video sequence is considered as a single attempt. The extracted probe/enrollment template is the result of a successful attempt and may contain a maximum of 60 (30 frames $\times$ 2 eyes) iris feature vectors. The successful construction of the feature vector means passing all the intermediate quality checks in the recognition pipeline.

The testing procedure consists of the following steps:

1. Passing of all the videos that satisfy the condition IN&NG through the feature extraction to produce the enrollment template: the template is considered as successfully created if the following requirements are met:
   a. At least 5 feature vectors were constructed for each eye;
   b. At least 20 out of 30 frames were processed.
2. Passing of all the videos through the feature extraction to produce the probe template, which is considered as successfully created in the case of at least 1 feature vector being constructed;
3. Creating of a pair-wise matching table of the dissimilarity scores for all the comparisons: each probe template is compared with all enrollment templates except the ones generated from the same video;
4. Calculating of the measures: FTE, FTA, FNMR(FMR) and FRR(FAR).

One important thing that makes the enrollment and verification different are the values of the following thresholds: (i) the normalized eye opening (NEO) value, described in [10], was set as 0.5 for the enrollment and 0.2 for the verification; (ii) the non-masked area of the iris (not occluded by any noise) was set as 0.4 and 0.29 for the enrollment and probe, respectively.

5.2.3 Performance Evaluation

The recognition accuracy results are presented in Table 6. The proposed feature extraction and matching method is compared with the one proposed in [10] as a part of the whole pipeline. The compared method is based on Gabor wavelets with an adaptive phase quantization technique (Gabor+AQ). Both methods were tested in three different verification environments: indoors without glasses (IN&NG), indoors with glasses (IN&G), and outdoors without glasses (OT&NG). The enrollment was always carried out only indoors without glasses and, for this reason, the value of FTE = 3.15 is the same for all the cases. The target FMR = 10^{-7} was set in every experiment.

Applying different matching rules was also investigated. The proposed multi-instance fusion showed advantages over the other compared rules (Table 7).

Table 6 Recognition accuracy in different verification conditions

		Verification condition		
Error rate, %	Method	IN&NG	IN&G	OT&NG
EER	Proposed	0.01	0.09	0.42
	Gabor + AQ [10]	0.10	0.35	3.15
FNMR	Proposed	0.48	5.52	10.1
	Gabor+AQ [10]	1.07	8.94	32.5
FTA	–	0.21	4.52	0.59

Table 7 Recognition accuracy for different matching rules

Error rate, %	Method	Fusion	Minimum	Consensus
EER	Proposed	0.01	0.21	0.21
	Gabor+AQ [10]	0.10	1.31	1.31
FNMR	Proposed	0.48	0.92	1.25
	Gabor+AQ [10]	1.07	3.17	4.20

The overall execution time for the whole pipeline was measured on a single core of Qualcomm Snapdragon 835 CPU and was 55 milliseconds, which is about 18 FPS real-time performance.

6 Conclusion

A novel approach to iris feature extraction and matching was proposed in this paper. It showed robustness to the high variability of the iris representation caused by changes in the environment and physiological features of the iris itself. The profitability of using shallow textural features, feature fusion, and variational inference as a regularization technique, was also investigated in the context of the iris recognition task. One more feature of the proposed solution is its multi-instance iris fusion, which helps to increase the performance in case the input image contains both eyes at the same time. The proposed solution was tested in the video-to-video scenario and showed its ability to work in real-time in an uncontrolled environment. Although it showed high accuracy indoors, the outdoor recognition is still challenging.

Acknowledgements This research was also supported in part by a research grant 19-07-01231 of Russian Foundation of Basic Research.

References

1. Meng W, Wong D, Furnell S, Zhou J (2015) Surveying the development of biometric user authentication on mobile phones. IEEE Commun Surv Tutorials 17(3):1268–12937
2. Rui Z, Yan Z (2018) Survey on biometric authentication: toward secure and privacy-preserving identification. IEEE Access 7:5994–6009
3. Das A, Galdi C, Han H, Ramachandra R, Dugelay J, Dantcheva A (2018) Recent advances in biometric technology for mobile devices. In: 2018 IEEE 9th international conference on biometrics theory, applications and systems (BTAS), pp 1–11. https://doi.org/10.1109/BTAS.2018.8698587
4. Li YH, Savvides M (2009) Iris recognition, overview. Encyclopedia of biometrics. Springer, New York, pp 810–819
5. Prabhakar S, Ivanisov A, Jain AK (2011) Biometric recognition: sensor characteristics and image quality. IEEE Instrum Meas Soc Mag 14(3):10–16
6. Information technology (2011) biometric data interchange formats – part 6: iris image data, annex b. ISO/IEC 19794-6:2011
7. Corcoran P, Bigioi P, Thavalengal S (2014) Feasibility and design considerations for an iris acquisition system for smartphones. In: 2014 IEEE fourth international conference on consumer electronics Berlin (ICCE-Berlin), pp 164–167. https://doi.org/10.1109/ICCE-Berlin.2014.7034328
8. Bowyer KW, Hollingsworth K, Flynn PJ (2008) Image understanding for iris biometrics: a survey. Comput Vis Image Underst 110(2):281–307. https://doi.org/10.1016/j.cviu.2007.08.005
9. Daugman J (2004) How iris recognition works. IEEE Trans Circuits Syst Video Technol 14(1):21–30. https://doi.org/10.1109/TCSVT.2003.818350
10. Odinokikh GA, Fartukov AM, Eremeev VA, Gnatyuk VS, Korobkin MV, Rychagov MN (2018) High-performance iris recognition for mobile platforms. Pattern Recognit Image Anal 28(3):516–524. https://doi.org/10.1134/S105466181803015X
11. Thavalengal S, Corcoran P (2016) User authentication on smartphones: focusing on iris biometrics. IEEE Consum Electron Mag 5(2):87–93. https://doi.org/10.1109/MCE.2016.2522018
12. Zhang M, Zhang Q, Sun Z, Zhou S, Ahmed NU (2016) The BTAS competition on mobile iris recognition. In: 2016 IEEE 8th international conference on biometrics theory, applications and systems (BTAS), pp 1–7. https://doi.org/10.1109/BTAS.2016.7791191
13. Zhang Q, Li H, Sun Z, Tan T (2018) Deep feature fusion for iris and periocular biometrics on mobile devices. IEEE Trans Inf Forensics Secur 13(11):2897–2912. https://doi.org/10.1109/TIFS.2018.2833033
14. Tabassi E (2011) Large scale iris image quality evaluation. In: Proceedings of international conference of the biometrics special interest group (BIOSIG), pp 173–184
15. Matveev I, Novik V, Litvinchev I (2018) Influence of degrading factors on the optimal spatial and spectral features of biometric templates. J Comput Sci 25:419–424
16. Fujitsu develops prototype smartphone with iris authentication. Press Release (2015). https://www.fujitsu.com/global/about/resources/news/press-releases/2015/0302-03.html
17. Callaham J (2015) Microsoft Lumia 950 and Lumia 950 XL smartphones officially announced. Windows Central. Mobile Nations. https://www.windowscentral.com/microsoft-lumia-950-and-lumia-950-xl-smartphones-officially-announced
18. How does the iris scanner work on Galaxy S9, Galaxy S9+, and Galaxy Note9? (2019) https://www.samsung.com/global/galaxy/what-is/iris-scanning/
19. Galaxy tab iris (sm-t116izkrins) specification (2019) https://www.samsung.com/in/business/tablets/galaxy-tab-iris-7-0-t116ir/sm-t116izkrins/
20. Irisguard EyePay Phone (IG-EP100) specification (2019) https://www.irisguard.com/node/57
21. Minaee S, Abdolrashidi A, Wang Y (2016) An experimental study of deep convolutional features for iris recognition. In: 2016 IEEE signal processing in medicine and biology symposium (SPMB), pp 1–6

22. Gangwar AK, Joshi A (2016) DeepIrisNet: deep iris representation with applications in iris recognition and cross-sensor iris recognition. In: ICIP 2016, Phoenix, 25–28 Sept 2016, pp 2301–2305. https://doi.org/10.1109/ICIP.2016.7532769
23. Tang X, Xie J, Li P (2017) Deep convolutional features for iris recognition. In: Chinese conference on biometric recognition, pp 391–400. Springer
24. Proença H, Neves JC (2017) IRINA: iris recognition (even) in inaccurately segmented data. In: 2017 IEEE conference on computer vision and pattern recognition, CVPR 2017, Honolulu, 21–26 July 2017, pp 6747–6756. https://doi.org/10.1109/CVPR.2017.714
25. Zhao Z, Kumar A (2017) Towards more accurate iris recognition using deeply learned spatially corresponding features. In: IEEE international conference on computer vision, ICCV 2017, Venice, 22–29 Oct 2017, pp 3829–3838. https://doi.org/10.1109/ICCV.2017.411
26. Abate AF, Barra S, D'Aniello F, Narducci F (2017) Two-tier image features clustering for iris recognition on mobile. In: Petrosino A, Loia V, Pedrycz W (eds) Fuzzy Logic and Soft Computing Applications. Lecture Notes in Artificial Intelligence, vol 10147, pp 260–269
27. Zeiler MD, Fergus R (2014) Visualizing and understanding convolutional networks. In: Proceedings of ECCV
28. Harley AW (2015) An interactive node-link visualization of convolutional neural networks. In: Proceedings of ISVC
29. Pavelyeva E, Krylov A (2010) An adaptive algorithm of iris image key points detection. In: 20th international conference on computer graphics and vision, GraphiCon'2010, pp 320–323
30. Howard A, Zhu M, Chen B, Kalenichenko D, Wang W, Weyand T, Andreetto M, Hartwig A (2017) Mobilenets: efficient convolutional neural networks for mobile vision applications. CoRR abs/1704.04861 . http://arxiv.org/abs/1704.04861
31. Koch G, Zemel R, Salakhutdinov R (2015) Siamese neural networks for one-shot image recognition. In: Proceedings of the 32th international conference on machine learning
32. Kingma DP, Salimans T, Welling M (2015) Variational dropout and the local reparameterization trick. In: Cortes C, Lawrence ND, Lee DD, Sugiyama M, Garnett R (eds) Advances in neural information processing systems, vol 28, pp 2575–2583. Curran Associates, Inc. http://papers.nips.cc/paper/5666-variational-dropout-and-the-local-reparameterization-trick.pdf
33. Ross A, Jain A, Nandakumar K (2006) Handbook of multibiometrics. Springer, New York
34. Matveev I, Novik V (2019) Using optimal circular path method to match piecewise iris templates. Pattern Recognit Image Anal 29(1):194–202
35. Dunstone T, Yager N (2009) Biometric system and data analysis: design, evaluation, and data mining. Springer, Boston
36. Chinese Academy of Sciences, Institute of Automation (CASIA), CASIA-Iris-Mobile-V1.0 (2015) http://biometrics.idealtest.org/

A Protocol for Decentralized Biometric-Based Self-Sovereign Identity Ecosystem

Asem Othman and John Callahan

Abstract Most user authentication methods and identity proving systems rely on centralized databases. Such information storage presents a single point of compromise from a security perspective. If this system is compromised, it poses a direct threat to a significant number of users' digital identities. A recent example of compromised data includes the Equifax breach, which affected 140 million people. The other issue with these centralized systems that individuals don't have a control of how much of their Personal Identifying information (PII) is shared in different contexts.

This chapter discusses a decentralized biometric-based authentication protocol for identity ecosystems, called the Horcrux (The term "Horcrux" comes from the Harry Potter book series in which the antagonist (Lord Voldemort) places copies of his soul into physical objects. Each object is scattered and/or hidden to disparate places around the world. He cannot be killed until all Horcruxes are found and destroyed.) protocol, in which there is no such single point of compromise. The Horcrux protocol is founded on the principle that an individual should have a control over the use of their own PII. The decentralization of control over the components of individual identities will allow them proof of their PII – secured by blockchains and cryptography – to governmental and private-sector entities. Meanwhile, BOPS will enable these entities to undertake an advanced risk assessment, verify identities and provide seamless access through secure mobile biometric recognition technology. All of this can be achieved without the need to store PII in one central database and pose too great a risk for stakeholders. Horcrux protocol relies on decentralized identifiers (DIDs) under development by the W3C Verifiable Claims Community Group and the concept of self-sovereign identity. In this chapter, we discuss the specification and implementation of a decentralized biometric credential storage option via blockchains using DIDs and DID documents within the IEEE 2410–2017 Biometric Open Protocol Standard (BOPS).

A. Othman (✉) · J. Callahan
Veridium IP Ltd, Boston, MA, USA
e-mail: aothman@veridiumid.com

T. Bourlai et al. (eds.), *Securing Social Identity in Mobile Platforms*, Advanced Sciences and Technologies for Security Applications,
https://doi.org/10.1007/978-3-030-39489-9_12

Keywords Blockchain · IEEE BOPS · Self-sovereign identity · Authentication factors · Digital identity · Distributed authentication architecture

1 Introduction

The mobility of applications and networks have made the access to our Personal Identifying Information (PII) widely distributed, but most security and privacy preservation schemes are still primarily based on archaic, static models that don't work anymore and it is getting worse. The latest evidence of this is recent breaches disclosed by Yahoo, Equifax [1, 13], and Target stores [7] that has exposed identity information for millions of individuals. Hacking attacks are not targeting only enterprises but also federal agencies, such as the stolen database of fingerprint images in the US Office of Personnel Management breach of 2015 [30], and like other PII that is stolen, the unauthorized access of biometric data can be quite damaging to an individual. Despite these breaches and attacks, enterprises and national governments continue to take on enormous risk by aggregating unnecessary personal data while customers cannot manage the massive number of IDs, passwords and PII required to interact with every online connection.

We believe that the common denominator across most aspects of Personal Identifying information (PII) protection is identity. An identity is inextricably linked to a person, device, application, system or network and it is the most dependable 'perimeter' we can rely upon to determine how to make information available securely and adequately. Meanwhile, user authentication is the security requirement in any identity ecosystem to access PII. User authentication can be described as a process in which a user offers some form of proof that s/he is the same user who registered the account. Proof of identity can be any piece of information that an authentication server accepts: something users have in their possession (e.g., tokens), something they know (e.g., username and password) or something they are (e.g., a biometric).

By reviewing current identity proving ecosystems (see Fig. 1), we can determine these systems rely on specific parties: an *issuer*, *end-user*, *verifier*, and *inspector*.

Issuers such as governments associate identity credentials to end-users. Then, the issuer shares personal information and credentials of the end-user with a verifier. If the end-user applies for a bank account, credit card, or car loan, the inspector contacts a verifier to prove the claimed identity by the end-user. Therefore, especially if this process is online, the inspector presents a multiple-choice quiz about past addresses or who financed the user's last car. That is an identity verification service that verifier provides to lenders and others, i.e., inspectors. Based on the answers or prove of holding the credentials, the inspector will verify the claimed identity by the end-user and grantee the required service.

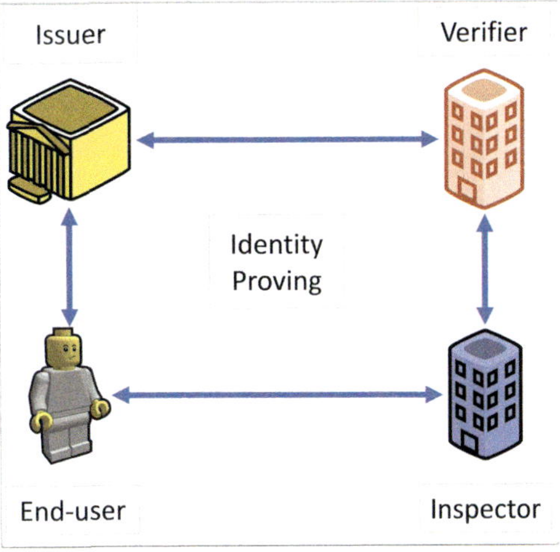

Fig. 1 Traditional identity proving ecosystem

In current digital and interconnected practice, these verifiers become a centralized database which stores the data used for authentication. When the user offers the requested proof of identity, the authentication server evaluates this proof and grants access to the user. For example, when a user tries to access his account on a typical web application, he is prompted to enter a password. Traditionally, the web application holds the information about the user's account and his password. When the user submits his password during the log-in process, the application compares the stored password to the submitted password. If they match, the user is granted access to the application. In other words, all the information needed to authenticate the user is held on a single system. Even if the authentication system is a biometric-based system, most of the deployed systems is still use the same centralized model.

In these traditional authentication and identity models, users are forced to relinquish personal information such as credit histories, credentials such as birth certificate, or biometric data such fingerprint template to a third party, with a centralized database. Moreover, users do not have their own consolidated digital identity; they have tens or hundreds of fragments of themselves scattered across different organizations, with no ability to control, update or secure them effectively. These security flaws encapsulate perfectly why identity became the new attack surface [16].

Hence, it is of paramount importance to facilitate an identity ecosystem that leverages personas, reduces liability for the enterprises, provides distributed access to authorized services, and provides the user full control over their identity accessing via a privacy-centric biometric-based authentication model.

In this chapter, we discuss the specification and implementation of our Horcrux protocol that combines the new decentralized self-sovereign identity ecosystem with 2410–2017 IEEE Biometric Open Protocol Standard (BOPS) [2].

Self-sovereign identity (SSI) is a new decentralized ecosystem for private and secure identity management that is being implemented by several projects [6, 17] as the replacement for traditional identity proving systems. Self-sovereign identity puts end-users – not the organizations that traditionally centralize identity – in charge of decisions about their privacy and disclosure of their personal information and credentials. Self-sovereign identity utilizes distributed ledgers (DLT), i.e., blockchain technology, to establish a web-of-trust [9].

Biometrics Open Protocol Standard, or BOPS, is an IEEE standard 2410–2017 [2]. BOPS supports a distributed storage model, which is neither device- nor server-centric storage [28], where the user's biometric template is distributed using a secret sharing scheme between the user's mobile device and the service provider. Both shares of the biometric data are encrypted, and for the authentication process to be successful, both shares are required [2].

Horcrux protocol utilizes SSI and BOPS to implement a secure and robust identity-authentication solution capable of supporting different business requirements as well as the privacy of users by allowing them to manage the storage and access of their Personal Identifying information (PII)[1] via a distributed mobile biometric authentication system. This marriage of these two models (SSI and BOPS) via the Horcrux protocol will guarantee the following principles:

- *Existence*: users must have an independent existence that can not only exist wholly in the digital form, and by using a biometric-based protocol, i.e., BOPS [2] for enrolling and authentication, this guarantees that the digital identity has been created and will always be verified by an existing end-user.
- *Control*: users must control the storage and access to their identities. Under the Self-sovereign identity ecosystem, users are always able to refer to, update, or even hide their personal information and credentials. The Horcrux protocol will assure that the access is always secure by their biometric which also is securely stored via the decentralized ecosystem, along with their personal information.
- *Portability and interoperability*: BOPS [2] and self-sovereign identity have been designed around these principles.
- *Protection*: the security of the Horcrux protocol is trusted because it is based on strong cryptography and governed by self-sovereign identity using a blockchain technology and BOPS.

The rest of the chapter is organized as follows. Section 2 gives a quick overview of the different identity models and evolution of these models into the new self-sovereign identity ecosystem that provides users with full control over their identity access and storage. In Sect. 3, we discuss the biometric authentication standard BOPS and the unique way to store biometric data in a distributed matter to preserve the privacy and security of the stored biometric data. Section 4 looks at the Horcorx protocol where both BOPS and SSI model can be deployed together to provide a

[1]Personal Identifying information are Data about an individual which considered to be sensitive and thus subject to security and privacy protections such as biometric and demographic data.

new way for users to establish a portable, secure and controllable biometric-based identity system which is intrinsically theirs. Finally, Sect. 5 summarizes the chapter.

2 Self-Sovereign Identity Ecosystem

The internet and online services were built without a standard, explicit way of identifying people or organizations. So websites simply began offering their own local accounts with usernames and passwords, and this has been the predominant solution ever since.

But this silo-based approach, where users must maintain identities for every site they interact with, has become untenable. It is not just a usability disaster for individuals, it also creates a multitude of data honeypots for hackers which when breached, compromises trust in all Internet services. At the same time, there is a growing economic inefficiency when organizations have to collect, store and protect the same sort of personal data in their own silos. It is reaching a tipping point.

To solve this problem, in some current implementations, the authentication server can be completely separated from the server running web applications or biometric authentication database. For example, single sign-on (SSO) schemes [24] are based on this concept. SSO schemes rely on a third-party identity provider (IdP) to broker authentication using protocols such as SAML [12] and OpenID Connect [29]. Since their introduction in 2002 and 2010 respectively, only 5% of sites use any of over 50 disparate IdP [32] SSO services (e.g., "login with Facebook", "login with Google", etc.).

However, these have produced inadvertent side effects such as concentrating control around a small number of providers, increasing data leakage through inadvertent sharing, and raising privacy concerns, all while not actually giving the individual real control.

Surveys of users show an overwhelming dissatisfaction with single-sign-on (SSO), a feeling of "lack of control" over their data [19, 27, 31] and a desire to control it themselves. Recent legislation, such as the General Data Protection Regulations (GDPR) [3, 15] and Payment Services Directive II (PSD2) [10], are pressuring institutions, both private and public, to place citizen or customer data into the end user's control.

Therefore, recently digital identity ecosystems are moving from centralized to a common identity layer that allows people, organizations and things to have their own self-sovereign identity—a digital identity they own and control, and which cannot be taken away from them.

Self-sovereign identity is a new identity ecosystem where individuals (or even organization) to whom the identity pertains, control and manage their identities. In this sense the individual is their own identity provider – no external party can claim to "provide" the identity for them because it is intrinsically theirs. In other words, self-sovereign identity is as a digital record or container of identity transactions that end-users control. The end-user can add more data to it, or ask others to do so, reveal some the data or all of it some of the time or all the time.

Moreover, end-users can record their consent to share data with others, and easily facilitate that sharing. It is persistent and not reliant on any single third party. Claims made about an end-user in identity transactions can be self-asserted or asserted by a 3rd party whose authenticity can be independently verified by a relying party. The infrastructure of self-sovereign identity has to reside in an environment of diffuse trust which is not controlled by any single organization or even a small group of organizations. The cryptographically secure blockchain is the breakthrough technology that makes this possible. It enables multiple entities such as organizations and governments to cooperate mutually via distributed consensus to form decentralized blockchains, where data is replicated in multiple locations to be resistant to faults and tampering. While distributed ledger technology has been around for some time, new blockchain applications, such as Bitcoin, have resulted in realizations of its potential, particularly with respect to decentralization and security.

Figure 2 provides an overview of the self-sovereign identity architecture. The followings are the brief descriptions of the architecture entities. Note that in this architecture, the information is no longer centralized and connections are individually permissioned.

- *DID:* Decentralized Identifiers (DIDs) are a new type of identifier intended for a self-sovereign identity system, i.e., entirely under the control of an entity and not dependent on a centralized registry or certificate authority. DIDs are opaque,

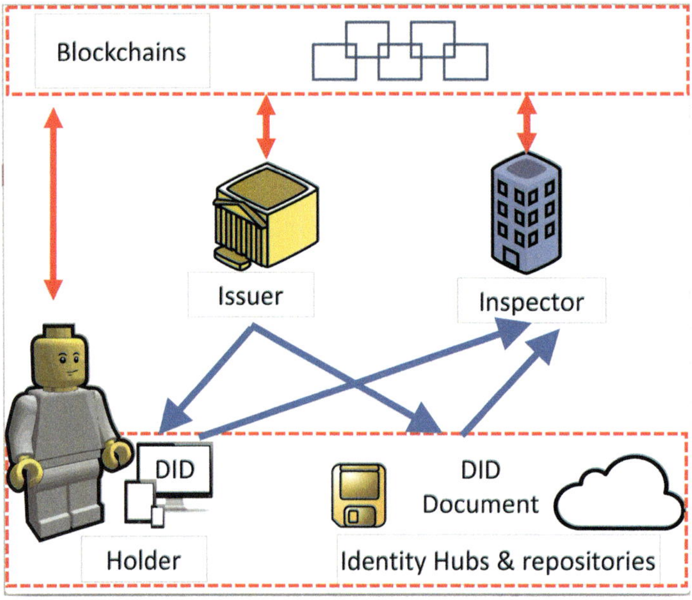

Fig. 2 Self-sovereign identity ecosystem architecture

unique sequences of bits, that get generated when a user accepts a claim from an issuer along with a corresponding DID Document. DIDs have a foundation in (Universal Resource Identifiers) URIs [18, 25]; therefore, they achieve global uniqueness without the need for a central registration authority.
- *DID document:* A DID resolves to a corresponding DID Document—a simple document that contains all the metadata needed to interact with the DID. Specifically, a DID Document typically contains at least three things along with personal information or credentials. The first is a set of mechanisms that may be used to authenticate as a particular DID (e.g., public keys, biometric templates, or even an encrypted share of biometric data). The second is a set of authorization information that outlines which entities may modify the DID Document. The third is a set of service endpoints, which may be used to initiate trusted interactions with an entity [25].
- *Blockchains:* In this architectural construct, the blockchain acts as an index of identifiers and audit trail of permissioned exchanges between the issuer of claims, the holder of claims, and the inspector of claims.
- *Identity hubs and repositories:* These hubs are secure personal data repositories that curate and coordinate the storage of signed/encrypted DID documents, and relay messages to identity-linked devices. Examples of identity hubs include Dropbox, Google drive, and Storj.
- *Issuer:* An entity that creates DID and DID documents, associates it with a particular subject and transmits it to a holder. Examples of issuers include corporations, governments, and individuals.
- *Inspector/Verifier:* Inspectors request claims in the form of DIDs from subjects and organizations in order to give them access to protected resources. The inspector verifies that the credentials provided via DID and in the DID document are fit-for-purpose, also checks the validity of the DID in the blockchain. Examples of inspectors include employers, security personnel, and websites.
- *Holder:* Holders receive DIDs from issuers, store DID Documents via identity hubs, and provide DID Documents to inspectors. The entity which controls a particular DID can be the subject of the DID document, but not necessarily. An inspector can also resolve DIDs into their corresponding DID documents and discover DIDs across a decentralized system. Examples of holders are users—students, employees, and customers. Other examples of holders that have the permissions to handle subject's claims include web services or mobile apps installed on the subject's personal devices.

SSI users have the liberty to manage their identity data on their mobile devices or cloud repositories. Mobile devices have become an essential part of our lives. We use mobile devices to store our credentials and payment. Therefore, while physical documents and storage of identity attributes on the cloud and third-party identity providers may exist for the years to come, storing identity data on mobile devices is the next natural step towards the realization of Self-Sovereign Identity and using mobile biometric can help in facilitating this to protect and authenticate digital identities.

Although mobile biometric authentication has the potential to offer significant value to enterprises, the unauthorized access of biometric data can be quite damaging to individuals due to its uniqueness and intrinsic to them. Currently, biometric data is stored in the mobile or server. These data storages are done within encryption layers including choosing a proper compliance system and infrastructure, which considers the particularly sensitive nature of biometric data. Nevertheless, with the news of stolen and hacked biometric data from phones [33], as well as servers breaches [30], means these schemes of storage are not the best solution.

3 IEEE Biometric Open Protocol Standard (BOPS) Storage Model

In traditional authentication systems such as password and PIN, only one centralized database stores the data used for authentication. When the user offers the requested proof of identity, the authentication server evaluates this proof and grants access to the user. While most security experts and enterprises see the benefits of biometric-based mobile authentication in comparison to knowledge-based systems (usually, password and PIN), the underlying architecture with which to implement biometrics is still the same centralized storage model[2]; more specifically, whether a server or mobile-centric storage approach. The following describes the server- and mobile-centric approaches. Then we describe the distributed storage model that has been adopted by IEEE BOPS.

3.1 Server-Centric Approach

In this setup, biometric identity data is captured by trusted means and then stored centrally on a secure server. The server-centric biometric authentication architecture is managed by the service provider. To perform a user verification, the captured biometric sample is sent to the server for processing and matching against the enrolled data stored centrally.

A server-centric approach is likely preferred for organizations that desire a high degree of control over the end-to-end process of biometric authentication and to manage and secure the storage and use of the biometric data.

This approach also supports users accessing digital services via a wide range of endpoints such as computers, mobile devices, smart TVs, and physical locations (bank branch, enterprise access control, and in-store retail scenarios). Organizations

[2]The enrollment stage of most of the deployed biometric systems generates a digital representation of an individual's biometric trait that is stored in the system storage database [14].

can also analyze the biometric data they collect to improve the performance of matching algorithms.

Finally, by storing more resources and functions in the cloud rather than on the device itself, it reduces app size and complexity. As a result, server-centric authentication may also function more effectively with devices that have limited memory and processing power.

Comments on server-centric approach:

The major concerns with this approach are security and privacy. A server-based biometric database becomes a "honeypot" target for criminals, hostile governments and hacking groups. As the 2015 OPM hack [30], which led to the theft of millions of United States government personnel fingerprint data, demonstrated storing peoples' biometric data in network accessible databases can lead to wide-scale theft of sensitive data. Furthermore, there is the privacy concern of function creep where the biometric data is used in different purposes than authentication such as improving matching algorithms, databases linkage without consent, and deriving additional demographic information [20, 23].

Moreover, it is a generally accepted privacy principle that individuals must be able to access their PII and update it where necessary; therefore, some jurisdictions have already specifically referenced biometric data in privacy guidance and legislation such as European General Data Protection Regulation (GDPR) [3].

GDPR is European Union's new set of policies on data protection that officially took effect on May of 2018. While this regulation focuses on the citizens of European Union (EU), and reshapes the way organization across Europe handle citizens' PII data, any organization outside of EU that collects or processes data of EU citizens is also affected. GDPR expressly identifies biometric data as a category of sensitive personal data and requires the development of solutions with adequate privacy measures in place giving individuals' choice and control of their data. This means that organizations must ensure that individuals can access their biometric data as and when they request it. Further, organizations must have processes in place to allow individuals to correct, update and delete their data where necessary.

Based on such data privacy regulation, compared to sever-centric storage of biometric data, the storage and matching of biometric data on smartphones for authentication purposes is a compelling and more straightforward approach to satisfy global privacy requirements.

3.2 Mobile-Centric Approach

In this setup, biometric template creation, storage, and matching all occur locally on the device which allows an organization such as a bank to enable strong biometric authentication into their mobile app without having to manage PII on a central server. The mobile-centric biometric systems are getting growing support for solutions which are incorporating FIDO authentication protocols [5]. In a FIDO-compliant system, a successful biometric match grants access to a private key stored on the device, which is in turn used to respond to a Public Key Infrastructure (PKI) challenge[3] [4] from a relying party, such as a bank or retailer whose app is running on the device.

A mobile-centric approach is likely the best option for organizations with a primary objective of preventing large-scale breaches of customer data and satisfying global privacy requirements. Storing and matching biometric data on a device gives users more control over their data.

The mobile-centric approach for storing biometric data is also gaining momentum because now most major smartphone manufacturers are shipping devices that support biometric authentication and providing access to third-parties via APIs. These advances are enabling organizations to swiftly roll-out mobile-based biometric authentication services. Therefore, this mobile-centric model is being adopted by organizations, including banks and payment service providers (PSPs), as a quick way of solving the "password" problem.

Comments on Mobile-centric approach:

The manufacturer-led, mobile-centric model only solves part of the problem of providing secure and convenient access. Organizations are still looking at alternatives to ensure that an authentication solution is available to a large percentage of their users base. The mobile-centric approach only offers biometric authentication to those equipped with the latest mobile devices with integrated biometric sensors and secure hardware to store sensitive biometric data. In addition mobile biometric apps are consuming more disk and runtime footprints since the biometric processes all take place on the app, which less powerful devices may not easily support.

Moreover, as the data remains on the device, there are no transfers of the biometric data unless users perform backups to the cloud to avoid re-enrolling in cases of lost or damaged devices. However, most of the organizations that adopt the mobile-centric approach do not provide such backup services.

Finally, there are genuine concerns for organizations operating in highly regulated sectors, such as finance and healthcare, that this model to capture and store biometric data is managed by smartphone manufacturers using algorithms tuned to be more convenient than secure.

Although most of these deployed mobile biometric authentication systems by manufacturers are applying mechanisms to protect the integrity and confidentiality of data storage and code execution (i.e., TrustedExecution Environments [11] and Secure Elements [26]), Zhang et al. [33] revealed some severe issues with one of the deployed Android fingerprint frameworks which is using an embedded fingerprint sensor. They exploited an HTC One device with malware and demonstrated that an attacker can collect fingerprint images of victims every time they swipe their fingers.

[3]The private key is used to respond to the PKI challenge and never leaves the mobile device.

3.3 BOPS Distributed Storage Approach

The choice of either a device- or server-centric biometric authentication method provides organizations with both positive and negative consequences. However, the main concern with both approaches that there is a single point to compromise biometric data.

There is, however, a third approach that is a privacy-centric and also provides service providers with a mechanism of managing the storage of their customers/employees data without relying only on the operating system provided by a device manufacturer. This model is a distributed storage model that has been introduced by Othman and Ross [23] and adopted by the Biometrics Open Protocol Standard, or BOPS, which is IEEE standard 2410–2017.

The IEEE 2410–2017 Biometrics Open Protocol Standard (BOPS) [2] demands high levels of assurance to control communication between an organization server and its clients via two-way secure socket layer/transport layer security (SSL/TLS) and to monitor authentication logs and patterns with enhanced intrusion detection system (IDS) analytics.

The difference between BOPS approach and the aforementioned approaches (server- or mobile-centric) that the biometric enrolled data, i.e., representation of a fingerprint, voice, facial features, is cryptographically protected into two shards using a secret sharing scheme, i.e., Visual Cryptography [21]. These encrypted shards are stored, respectively, on a client device and a remote BOPS server, such that the biometric data is not kept in a single point to compromise.

Visual Cryptography Scheme [21] (VCS) is a simple and secure way to share a secret such that decryption can be performed using a simple binary operation. The basic scheme is referred to as the k-out-of-n visual cryptography scheme which is denoted as (k, n) VCS [21]. Given an original binary data T, it is encrypted into shares, such that:

$$T = S_{h_1} \oplus S_{h_2} \oplus S_{h_3} \oplus \ldots \oplus S_{h_k} \tag{1}$$

where $\oplus$ is a boolean operation, S_{h_i}, $h_i \in 1, 2, \ldots, k$ is a share which appears as white noise image, $k \leq n$, and n is the number of these shares. It is difficult to decipher the secret T using individual S_{h_i}'s [21]. The encryption is undertaken in such a way that k or more out of the n generated shares are necessary for reconstructing the original secret T.

As shown in Fig. 3, BOPS defines three steps during enrollment. First, the remote server generates a public-private key pair (RKP) in which the public key is sent to the mobile device. Then, a biometric template (called the initial biometric vector or "IBV") is collected, encrypted into two shares (shard I and II) using 2-out-of-2 scheme, and then paired with a device-generated public-private key pair (LKP). In the third step, the LKP private key is reserved locally and the LKP public key along with the biometric share II are encrypted with the RKP public key for transmission to the server over a two-way TLS connection and IBV is discarded. The client

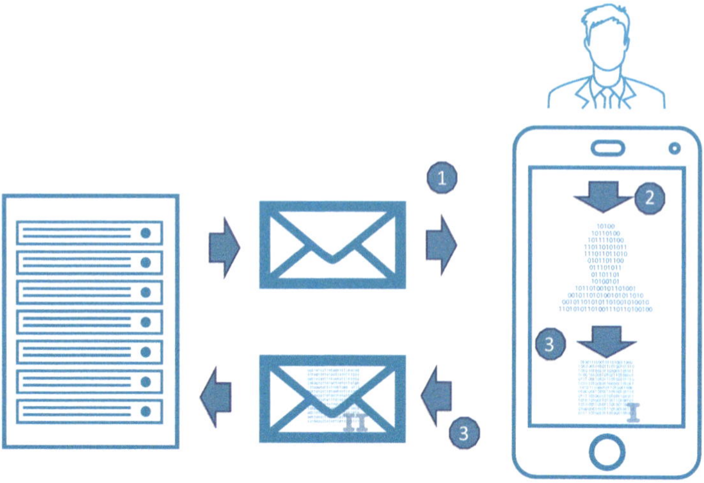

Fig. 3 Illustration of distributed model steps during the biometric enrollment stage. (1) Server sends an enrollment request along with the RKP public key, (2) Biometric capture is encrypted into two shares (I and II) using visual cryptography, and (3) Biometric share II and device (LKP) public key are encrypted by the server (RKP) public key and send to the server via two-way TLS

certificate for the TLS connection is installed a priori via application installation on the mobile device.

During authentication, a candidate biometric vector (CBV) is acquired for matching with IBV. BOPS defines two configuration modes for authentication:

- *Local Match*: The server is requested to encrypt (using its RKP private key) IBV share II it holds and returns them to the local device. The CBV is collected, IBV shares from local (I) and remote (II) combined and matched on the local device. The CBV and combined IBV are subsequently wiped from volatile memory.
- *Remote Match*: The collected CBV and the local IBV share I are encrypted in an envelope with the RKP public key and transmitted to the server. On the server, the incoming IBV share from the local device is combined with server-based share and compared to the incoming CBV. The CBV and combined IBV are subsequently discarded.

A distributed storage approach combines convenience, personal privacy, and enhanced security to create a model that makes it harder for attackers to compromise a system.

The fundamental idea of this distributed approach is utilizing secret sharing scheme [21] that, rather than encrypting the data as a single file using the standard public and private key pairing methodology, biometric data is encrypted randomly into multiple shares. These shares must be combined in order to recreate the original biometric data, ensuring that only the people, or devices, that possess the encrypted share files are able to recombine them and gain access to the protected information without any influence to the overall matching performance. Therefore, in the BOPS model, if the central biometric database, i.e., server is hacked, then attackers still need to have the user device's share of the biometric vector to break the system. Conversely, if a user has their mobile device compromised, an attacker still needs to break into the central database. This ensures that the biometric data is protected from data breaches, provides peace of mind for the end user that their biometric cannot be easily compromised, and enhances the storage architecture to eliminate misuse of the data. Moreover, this distributed model has two different matching configurations which allow an organization to customize their solutions based on their customer-base used technologies and network connectivity.

Hence, this simple IEEE open protocol standard solves the single point of failure and control concerns with a storage model that can lead to the deployment of more secure, flexible, and interoperable biometric authentication solutions.

In the following section, we discuss our Horcorx protocol where both BOPS and SSI model can be deployed together to provide a new way for users to establish a portable, secure and controllable biometric-based identity system which is intrinsically theirs.

4 The Horcrux Protocol

The IEEE 2410–2017 standard allows for interoperablility at several layers including the persistence cluster ([2] section 7.3.3) provided it satisfies security requirements for storage of encrypted biometric shares. We propose any BOPS server can act as a *holder* of biometric shares via blockchain using methods outlined in the W3C Decentralized Identity (DID) specification [25]. A BOPS server can enroll a user by storing biometric share(s) as DID Documents using off-chain storage providers owned by the user. The corresponding DID acts as the identity assertion associated with the enrolled biometric. Figure 4 depicts a standard BOPS enrollment flow (adapted from [2] section 7.2). The user (via a browser user-agent) is prompted to enroll their biometrics with a service provider acting as an *issuer*. The initial biometric vector (IBV) is encrypted (via visual cryptography) into two shares. One share is reserved on the local mobile device while the second is transmitted to the BOPS server. Instead of an RDBMS or persistence cluster (e.g., SOLR) backend, the BOPS server relies on a blockchain store in this case using a decentralized identifier (DID) [25] for persistence. DIDs provide a blockchain-agnostic method for resolving DID Documents much like URIs [18] uniquely characterize web resources via URNs and URLs, but for disparate blockchain ecosystems. The W3C Verifiable Claims Community Working Group has defined DID method specifications [25] for implementors of CRUD operations specific to a particular blockchain. The BOPS server acts as a resolver given a DID to fetch the corresponding DID Document if possible. The DID and corresponding DID Document are cryptographically associated with each other via blockchain transactions such that any tampering with the DID Document for a given DID would be evident. After persisting the DID document and registering the associated DID

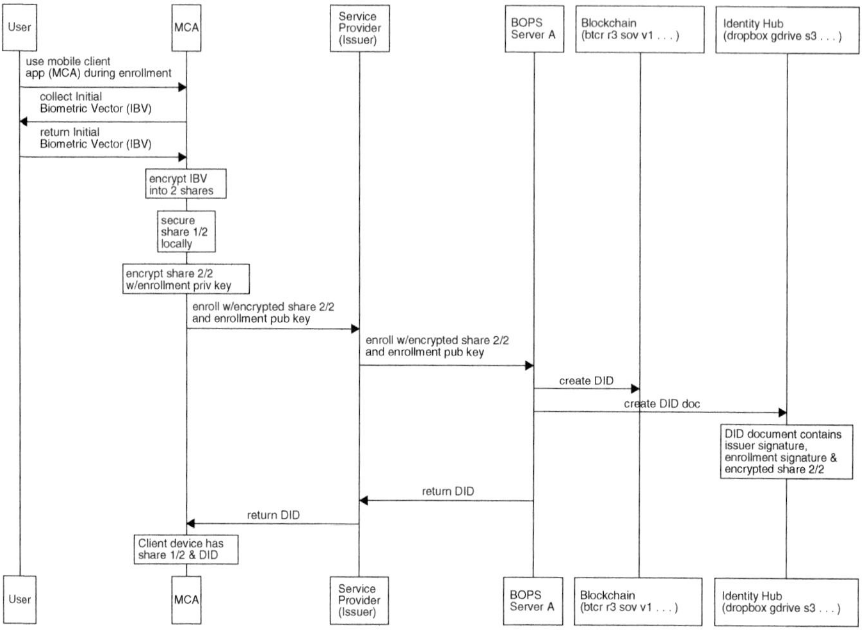

Fig. 4 Enrollment sequence

on a blockchain, the user is notified of success (or failure) of their enrollment. It should be noted that no biometric shares are stored on any blockchains, only in DID Documents that are persisted "off-chain" via identity hubs or personal storage providers.

The encrypted biometric share is still within an encrypted envelope as per [2] but the share is persisted on a corresponding blockchain with an associated DID. The DID can be used as a claim with another BOPS server acting as a *verifier*. Again, this is possible because any tampering with the DID Document associated with a given DID will be detectable due to their relationship via a recorded blockchain transaction [25]. Figure 5 shows an example of a different BOPS server being used by a verifier. In this example, the user tries to access a resource on a web site (e.g., the service provider) using a mobile client application (MCA) with a DID created by an issuer (4) and a public key created at enrollment. The service provider relies on a BOPS server to resolve the DID and fetch the corresponding DID Document via a blockchain from the storage provider. If the DID document is a valid claim, the BOPS server checks if the issuer of the claim is known (via its public key in the DID document) and that the enrollment public key matches for this user as well. If valid, the user (via their MCA) is requested for their candidate biometric vector (CBV) and complement share of the IBV as per [2]. Upon receiving the complementary share and CBV from the client (as described in 3 – Remote configuration mode), the enrollment public key is used to decrypt the client's share, combine the IBV shares and match them to the CBV. If successful, the user is authenticated.

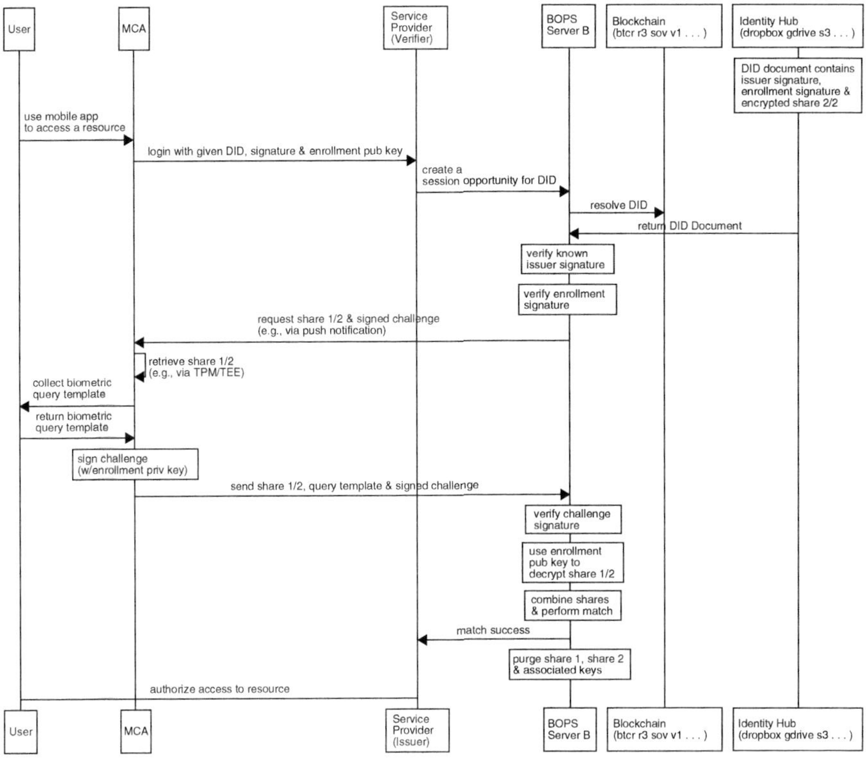

Fig. 5 Remote authentication sequence

In the case of remote authentication, the service provider, acting as a verifier, uses a different BOPS server instance to authenticate the user even though this user has never registered at this service provider. Furthermore, the user and service provider are the only parties needed at authentication time unlike SAML or OAuth that rely on 3rd party identity providers (IdPs) to broker identity claims in traditional single-sign-on (SSO) systems. The Horcrux protocol supports *self-sovereign identity* [8] by using blockchain technology to secure credentials issued by valid authorities (i.e., *issuers*) for later use directly by the user who owns the credentials. The user may store such credentials via several personal cloud storage providers such as Dropbox, Google drive, Amazon S3, etc. but delegate management (via OAuth tokens) to a *holder* such as the BOPS server. The holder can access issued claims like the encrypted biometric shares on behalf of the user during authentication, but require biometric authentication as specified in the `authenticationCredentials` section of the claim [25].

The local configuration mode of BOPS is also available such that a combination of biometric shares occurs on the mobile device. Figure 6 shows this variation in which the second biometric share is retrieved via DID referencing from the corresponding DID document but is transmitted to the client by a service provider

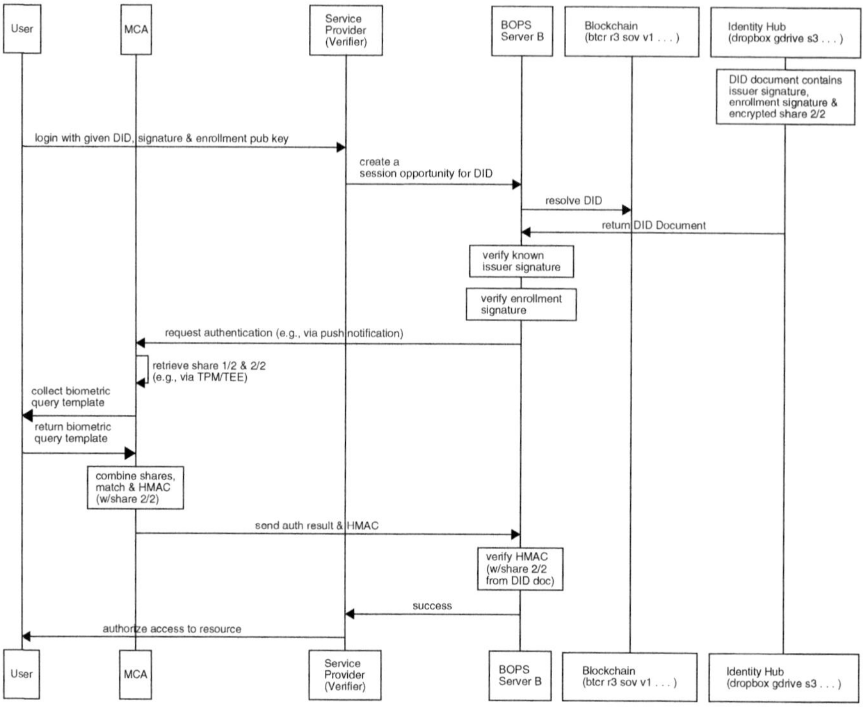

Fig. 6 Local authentication sequence

and its BOPS server. The biometric share is opaque to the service provider and BOPS server in this case, but the server knows that the corresponding share on the mobile device is used for matching due to the HMAC of the encrypted second share. The enrolled share is never sent to the device, but both shares are kept locally as per BOPS local configuration mode. The mobile device must hold the private key associated with the enrolled share for the DID because it computes an HMAC using the share and sends it to the server. The server can compare the HMAC key with the opaque encrypted share from the DID document. It is possible, however, that the user could resolved a given DID, retrieve the corresponding DID document, extract the opaque encrypted share and construct the HMAC thus spoofing possession of that share and falsifying the biometric match. We are in the process of investigating methods for securing DIDs on a mobile device and/or using server-based key mechanism to prevent this attack vector.

The IEEE 2410–2017 standard allows for more than two encrypted shares. Algorithms such as visual cryptography [28] and, Naor and Shamir secret sharing [21] allow for larger number of shares. Using DIDs and associated DID documents for more biometric shares across different blockchains and replicating copies of shares could further protect users from compromise and increase availability.

5 Summary

The threat of cyberattacks and the explosive growth of mobile and connected devices has ignited the quest for practical, secure and privacy-preserving digital identity and access management (IdM) architectures with highly secure authentication solutions.

While the Self-Sovereign Identity (SSI) model is the next evolution of identity management paradigm in which users have complete ownership and control over their digital identity, there is the need to provide the users with a secure, reliable and interpretable biometric authentication model to control the storage and access to their digital identities. The Horcrux protocol is a method for secure exchange of biometric credentials within an existing standard (IEEE 2410–2017 BOPS [2]) implemented across next-generation blockchain-based self-sovereign identity platforms based on open standards like DIDs and DID Documents [25]. By using blockchain and off-chain storage as an alternative to the persistent layer in BOPS, we use new blockchain-agnostic standards to enroll via an issuer and authenticate on a verifier that is not part of a real-time trust network. Instead, they rely on user-controlled biometric credentials that are cryptographically encrypted into multiple shares across the user's device and blockchain-linked personal storage providers. The protocol is generalized for two or more biometric shares that can be stored across mobile devices and personal storage providers with redundancy for availability and safety. Future plans include a reference implementation and detailed analysis of the protocol for performance and correctness using TLA+ in a manner similar to the protocol analysis of WPA found in [22].

Acknowledgements The authors would like to thank Ward Rosenberry for his help in editing and proofreading the chapter.

References

1. https://www.ftc.gov/equifax-data-breach (2020)
2. 2410–2017 IEEE biometric open protocol standard (BOPS). https://standards.ieee.org/findstds/standard/2410-2017.html
3. European union general data protection regulation (GDPR) (2020) https://gdpr-info.eu/
4. Public key infrastructure (2020) https://en.wikipedia.org/wiki/Public_key_infrastructure
5. FIDO UAF Protocol Specification v1.0 Proposed Standard (2014) https://fidoalliance.org/specs/fido-uaf-v1.0-ps-20141208/fido-uaf-protocol-v1.0-ps-20141208.html
6. Ali M, Nelson JC, Shea R, Freedman MJ (2016) Blockstack: a global naming and storage system secured by blockchains. In: USENIX annual technical conference, pp 181–194
7. Armerding T (2018) The 17 biggest data breaches of the 21st century. https://www.csoonline.com/article/2130877/data-breach/the-biggest-data-breaches-of-the-21st-century.html
8. Baars D (2016) Towards self-sovereign identity using blockchain technology. Master's thesis, University of Twente
9. Caronni G (2000) Walking the web of trust. In: IEEE 9th international workshops on enabling technologies: infrastructure for collaborative enterprises (WET ICE 2000). Proceedings. IEEE, pp 153–158

10. Cortet M, Rijks T, Nijland S (2016) PSD2: the digital transformation accelerator for banks. J Paym Strateg Syst 10(1):13–27
11. Ekberg JE, Kostiainen K, Asokan N (2013) Trusted execution environments on mobile devices. In: Proceedings of the 2013 ACM SIGSAC conference on Computer & communications security. ACM, pp 1497–1498
12. Hughes J, Maler E (2005) Security assertion markup language (saml) v2. 0 technical overview. OASIS SSTC Working Draft sstc-saml-tech-overview-2.0-draft-08, pp 29–38
13. Hume M (2018) Identity theft cited as threat after equifax security breach. The Globe and Mail, Toronto A, 7
14. Jain A, Ross A, Prabhakar S (2004) An introduction to biometric recognition. IEEE Trans Circuits Syst Video Technol 14(1):4–20
15. Koops BJ, Leenes R (2014) Privacy regulation cannot be hardcoded. Intl Rev Law Comput Technol 28(2):159–171
16. Los R (2016) The emergence of identity as an enterprise attack surface. CSO Online
17. Lundkvist C, Heck R, Torstensson J, Mitton Z, Sena M (2016) Uport: a platform for self-sovereign identity
18. Mealling M, Denenberg R (2002) Report from the joint W3C/IETF URI planning interest group: uniform resource identifiers (URIs), URLs, and uniform resource names (URNs): Clarifications and recommendations. Technical report
19. Mertens W, Rosemann M (2015) Digital identity 3.0: The platform for people
20. Nagar A, Nandakumar K, Jain A (2010) Biometric template transformation: a security analysis. In: Proceedings of the of SPIE, electronic imaging, media forensics and security XII, San Jose
21. Naor M, Shamir A (1994) Visual cryptography. In: Workshop on the theory and application of cryptographic techniques. Springer, pp 1–12
22. Narayana P, Chen R, Zhao Y, Chen Y, Fu Z, Zhou H (2006) Automatic vulnerability checking of IEEE 802.16 wimax protocols through TLA+. In: 2nd IEEE workshop on secure network protocols, 2006. IEEE, pp 44–49
23. Othman A, Ross A (2015) De-identifying biometric images by decomposition and mixing. In: Ngo D, Teoh A, Hu J (eds) Biometric security. Cambridge Scholars Publishing, Newcastle upon Tyne
24. Radha V, Reddy HD (2012)s A survey on single sign-on techniques. Procedia Technol 4:134–139
25. Reed D, Sporny M (2017) W3C decentralized identifiers (dids) 1.0. https://w3c-ccg.github.io/did-spec/
26. Reveilhac M, Pasquet M (2009) Promising secure element alternatives for NFC technology. In: First international workshop on near field communication, 2009. NFC'09. IEEE, pp 75–80
27. Rose J, Rehse O, Röber B (2012) The value of our digital identity. The Boston Consulting Group, New York
28. Ross A, Othman A (2011) Visual cryptography for biometric privacy. IEEE Trans Inf Forensics Secur 6(1):70–81
29. Sakimura N, Bradley J, Jones M, Medeiros B, Jay E (2011) Openid connect standard 1.0. [online] http://openid.net/specs/openid-connect-standard-1_0-21.html. Accessed 30 Mar 2013
30. Sanger DE (2015) Hackers took fingerprints of 5.6 million U.S. workers, government says. The New York Times
31. Satchell C, Shanks G, Howard S, Murphy J (2011) Identity crisis: user perspectives on multiplicity and control in federated identity management. Behav Inform Technol 30(1):51–62
32. Vapen A, Carlsson N, Mahanti A, Shahmehri N (2016) A look at the third-party identity management landscape. IEEE Internet Comput 20(2):18–25
33. Zhang Y, Chen Z, Xue H, Wei T (2015) Fingerprints on mobile devices: abusing and leaking. In: Black hat conference, Las Vegas

Towards Wider Adoption of Continuous Authentication on Mobile Devices

Sanka Rasnayaka and Terence Sim

Abstract Continuous Authentication (CA) is the process of constantly checking for the authorized user's presence, which brings unique advantages and disadvantages. CA is more secure and facilitates schemes with multiple levels of authentication security; however, it can consume more resources and cause user anxiety about privacy. In this chapter we seek to understand the practical aspects of CA; in particular, user perception and resource consumption. To gauge user perception towards CA, we conducted a survey with roughly 500 respondents. We found that users desire multiple levels of authentication security. Furthermore, users are willing to adopt CA for mobile devices. We then analyzed factors like security awareness, gender, and mobile device OS, to draw statistically significant conclusions regarding their effect on users' willingness to adopt CA, and user perceptions about CA. We also compare between biometric modalities based on their resource consumption, as measured by their Resource Profile Curve (RPC). This Curve reveals the trade-off between authentication accuracy and resource usage, and is helpful for different usage scenarios in which a CA system needs to operate. In particular, we explain how a CA system can intelligently switch between RPCs to conserve battery power, memory usage, or to maximize authentication accuracy. We argue for the importance of understanding user perceptions and using RPCs to guide the development of practical CA systems.

S. Rasnayaka (✉) · T. Sim
School of Computing, National University of Singapore, Singapore, Singapore
e-mail: sanka@comp.nus.edu.sg; sanka@u.nus.edu; tsim@comp.nus.edu.sg

T. Bourlai et al. (eds.), *Securing Social Identity in Mobile Platforms*, Advanced Sciences and Technologies for Security Applications,
https://doi.org/10.1007/978-3-030-39489-9_13

1 Introduction

> **Prolog:** Bob meets Alice on his way to work. Bob unlocks and gives his phone to Alice to show her the new game he installed yesterday. After playing the game for a while, Alice tries to open Facebook in Bob's phone while he is not looking. Since the phone is unlocked, Alice can go through Bob's social media profile. What can Bob do to better secure his mobile device?

With the rapid increase of mobile phone usage in day-to-day activities, including banking and e-commerce applications, the security requirement of these devices has increased drastically. However, this increased security requirement is not met by current authentication schemes available. The predominant authentication method currently used is one-time, session-based authentication, in which the user produces a secret known only to him (e.g. PIN, Password), or some form of biometrics (e.g. Fingerprint, Face) to authenticate himself before using the computing device.

This session-based authentication scheme was developed for desktop environments where, (i) the device would be physically near the user only while he is using it; (ii) each session would last for a long time; (iii) there would be few (one or two) such long sessions during the day; and (iv) the device would not be easily shared. However adopting the same authentication method on mobile devices raises many issues due to the inherent differences of usage between desktops and mobile devices. These differences are illustrated in Fig. 1.

Sessions on mobile devices tend to be very short and there would be many of these during a day (e.g. a few seconds to scroll through notifications, or to check Facebook). Therefore the time taken to authenticate at the beginning of each short session is a considerable overhead. Since mobile devices are compact, they tend to be more vulnerable to theft. Furthermore, mobile devices are readily shared, making them more vulnerable to exposing private information to family or friends.

	(i). Close Physical Proximity	(ii). Session Duration	(iii). Number of Sessions	(iv). Sharing
	Only when in use	Long	Few	Rare
	Always	Short	Many	Often

Fig. 1 Differences between Desktop and Mobile environments

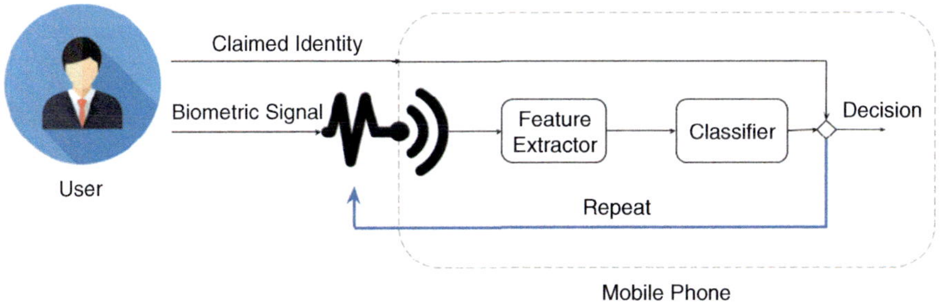

Fig. 2 Process of a Continuous Authentication system

Continuous Authentication(CA) is gaining traction as a better alternative to session-based authentication for mobile devices. A CA system works by monitoring the user of a system transparently using biometrics. The usual process of a CA system is depicted in Fig. 2. CA computes a confidence value that reflects the presence of the logged-in user at any given time. This confidence value allows the phone to enforce different security levels for different apps. For an example, a banking app would require a higher level of security when compared with a clock app to tell the time.

Continuous Authentication addresses the drawbacks of one-time authentication schemes highlighted above. With the use of biometrics the authentication process could be automated requiring little to no user involvement. Therefore the overhead of unlocking multiple times for the larger number of shorter sessions throughout the day is reduced. The mobile phone can continually monitor the authorized user because the device will be physically closer to the user even when it is unused. This will allow the device to notice as soon as the phone is shared with someone else, in which case access to sensitive data can be denied. This shows how CA can be useful in the scenario given in the Prolog at the beginning of the chapter. Refer to the Epilog for more details.

However, compared to desktop PCs, mobile devices are resource-constrained. They have limited amount of energy, computational capabilities, and memory. Continuously checking biometric signals would be an additional strain on these limited resources. Therefore CA researchers should be more aware of how different biometrics consume resources and how to effectively manage these resources.

As with any new technology dealing with security and privacy, widespread adoption will depend on how the end users perceive the technology, as well as how practical implementations of the technology can be incorporated into current hardware. Therefore this chapter looks into two aspects which are critical for the successful adoption of CA systems on mobile devices, namely, User Acceptance, and Resource Consumption.

1.1 User Acceptance

We analyze how people perceive Continuous Authentication, by introducing the concept using a scenario-based video and textual descriptions.

After introducing the respondents to CA, we analyze their perceptions toward biometrics, multi-level authentication and Continuous Authentication. The main objective is to identify the key concerns users have when moving to a new authentication method. Another objective is to understand the different factors influencing their perceptions. This understanding will then help software developers better design and implement CA, leading to greater user acceptance.

1.2 Resource Consumption

In a resource-limited environment like mobile devices, any utility provided should be measured with respect to the resources it consumes. The different levels of authentication security that each biometric can provide, and the resources used to provide them, can be characterized by a Resource Profile Curve (RPC). This paper is the first in the research literature for such an analysis.

The proposed Resource Profile Curves will have real-life implications for CA implementations. The following scenarios explain how RPCs will be useful.

- **Security-First Scenario:** Each app can specify a minimum level of authentication security before it can be started. (e.g. a banking app can specify a minimum level of 95%, while a clock app can specify a lower minimum of 10%.) RPCs can then be used to select the biometric modality (or combination of modalities) that achieves this minimum before launching the app.
- **Resource-First Scenario:** The user may choose to conserve battery power when his battery level is low, and he prioritizes phone calls over other apps. RPCs can be used to restrict what apps can be launched, because the app's minimum authentication level requires too much power to achieve (or warn the user of this).
- **Context-Based Scenario:** Depending on what activity the user is currently doing (e.g. walking and talking, or standing still and texting), some biometric modalities might not be available. For example, when the user is walking and talking on the phone, his facial image is not available, but his voice and gait are. The opposite is true when the user is standing still and texting. RPCs can be used to choose the available biometric modalities that can achieve the minimum authentication level for any app.

By understanding how each biometric modality performs within resource constraints a CA system can provide the highest possible security while maintaining the lowest possible resource consumption.

2 Related Work

2.1 Continuous Authentication

Many researchers have introduced Continuous Authentication schemes using various biometrics. Early attempts at CA used keystroke dynamics [3] and mouse dynamics [37] for desktops computers.

As different biometric modalities became available, they were adopted for CA systems as well. In the research literature, we can find CA systems that use face, fingerprint [33], gait [27], bioAura [25] touch gesture [14, 15], soft biometrics [26] and behavioural biometrics [10]. In addition, there are works that fuse multiple biometric modalities to achieve better authentication accuracy [21, 35].

Despite all these research, actual commercial implementations of CA systems for mobile devices are rare: as of this writing, UnifyID (https://unify.id) is the only commercial vendor that claims to provide CA for mobile phones. This dearth of commercial offerings could be due to the lack of understanding of user perceptions of CA. We hope the research presented here will help address this issue.

2.2 User Perceptions

Mobile phone authentication. Ben-Usher et al. [1] validated that every mobile phone user want their device to be secure. Authors also found that users perceive the security provided by PIN to be neither adequate nor convenient. Therefore a more secure and convenient unlock method is needed for mobile devices. This finding has been corroborated by other researchers in [7] and [4].

Biometrics. There have been several user surveys asking the respondents to rank different biometric modalities based on their percieved security. Deane et al. carried out the survey in 1995 [12], Seiger et al. [32] did a similar survey in 2010 and our survey includes a similar question in [29] which was done in 2018. A comparison of these three surveys revealed that secrets-based authentication schemes (like PIN) are now perceived as less secure when compared to biometric based authentication methods (like fingerprint).

Casanova et al. [17] carried out a survey and ranked different biometrics like voice, face and hand biometrics with respect to comfort of use and how secure they are perceived to be. Since Continuous Authentication will use biometrics, similar perceptions are applicable.

Device sharing. A key difference between desktops and mobile devices is that the latter is frequently shared. This creates unique security and privacy issues in mobile devices which are not present in the desktop setting.

Karlson et al. [18] studied the willingness of users to share the phone along with the security and privacy concerns of sharing mobile phones. The authors highlight the requirement of multiple levels of security for different applications rather than the single level security model used today.

Mattewes et al. [22] have studied the device-sharing dynamics. The authors found that trust among the shares and user convenience were the highest influencing factors for device sharing.

CA is presented as a viable solution to overcome security issues arising from device sharing.

Multiple levels of security. Seiger et al. [32] propose the idea of graded security. The focus here was to get a mapping between the biometric methods and levels of security needed. Researchers concluded that using one biometric for all security levels was preferred, rather than using different biometrics for the different levels of security.

This finding aligns within the context of one-time authentication, since having to use different methods depending on what you want to do is very inconvenient. However it is interesting to see how people would perceive multi-level security in the context of CA, which we will focus upon.

Different factors influencing security perceptions have been studied extensively in many previous work [2, 17, 24, 31]. Similar kind of analysis with respect to security awareness, age, gender, current mobile operating system (android vs iOS) will be carried out in the work presented in this paper.

Continuous Authentication Clarke et al. [8] developed a prototype CA system and evaluated the convenience and the intrusiveness with a control study of 27 people. This work gives a stepping stone to understanding how people perceive CA.

Khan et al. [19] carried out a more comprehensive analysis on the usability, convenience/annoyance and security of CA. Throughout the controlled lab study and survey of 37 respondents these factors were analyzed using Likert scale questions.

Our work builds upon these two studies by (1) Having a larger sample size and backing up our hypotheses with statistical analysis. (2) Analyzing different factors affecting the perceptions. Specifically analyzing the impact of security awareness towards the perceptions. (3) Providing design considerations for a CA system in order for it to be acceptable to users.

2.3 Resource Consumption

Since mobile devices are resource-constrained, especially in energy and memory, there have been many studies on analyzing and profiling these resources for different aspects (e.g. apps, embedded software etc.). In [28] Qian et al. did a resource usage profiling for mobile devices in different layers (transport layer, application layer etc.) of a mobile device and proposed a resource optimizer. In [13] Falaki et al. proposed a smartphone resource usage monitoring tool which measures usage context (CPU and memory) for research deployments. A similar usage measurement tool was proposed by Wagner et al. in [38] which collects usage based information from Android smart phones and quantifies resource usage by the collaborators.

Carroll et al. carried out a direct approach to measure the significance of energy drawn by components in a smartphone in [5], where they have analyzed the energy consumption as well as battery lifetime for usage patterns. Tiwari et al. in their work [36] proposed a power analysis technique which has been applied to two commercial microprocessors for embedded software. In the earlier stages of smart phones, researchers from Nokia came up with a software profiling tool [11] that could be used by developers to measure the power consumption of their applications.

Even though several analyses have been done on the usability and security of CA [9, 20, 29], no work has been done on how CA may negatively impact a device with limited resources. Our work tries to address this gap and provides a new dimension to answer the practical question: How should a CA system choose between different biometric modalities and algorithms to achieve good accuracy while reducing energy and memory consumption?

We explore this in Chap. 4; but let us begin with User Perceptions.

3 User Perceptions

3.1 Methodology: Survey Design

The goal of this study is to identify user perceptions towards biometrics, Continuous Authentication and multilevel security. Our survey was designed to test the following hypotheses,

Security Awareness

- H1. Gender has an impact on security awareness
- H2. Age has an impact on security awareness
- H3. Occupation has an impact on security awareness
- H4. Education level has an impact on security awareness
- H5. Mobile device OS has an impact on security awareness

Perceptions towards mobile phone authentication

- H6. Current unlock methods are perceived to be inconvenient
- H7. There is a perceived requirement of different levels of security
- H8. The perceived requirement of different levels of security depends on security awareness

Perceptions towards Continuous Authentication

- H9. Users are willing to use Continuous Authentication for mobile devices
- H10. Security awareness has an impact on the willingness to use CA

- H11. Convenience, Security, Transparency and Interruptions introduced by Continuous Authentication is perceived differently based on gender, current mobile device OS and security awareness

The survey was designed to evaluate above hypotheses with three sections,

Section 1: Demographics

First section of the survey focused on getting demographic information of the respondents including age, gender, education level and occupation, current mobile device and unlock methods. A 5 point Likert scale was employed to evaluate the convenience of the current unlock method. Finally a free text question was given for them to express any issues or concerns they have regarding the current unlock method.

Section 2: Biometrics and CA

At the beginning of this section a demonstration video that explained the basic concept of Continuous Authentication and how it would work using a scenario based animation was shown. (link to video: https://youtu.be/ksSCWuUB6Ps).

After watching the video, respondents were asked to answer questions about the requirement of different levels of security and Continuous Authentication.

- Yes/no questions were used to evaluate requirements of different levels of security and the willingness to use CA
- 5 point Likert scale questions were used to evaluate perceptions toward convenience, security, transparency and interruptions in CA
- 3 ranking activities were used to evaluate, (1) perception of security of different biometrics, (2) different apps according to security requirement and (3) different factors considered when selecting a new authentication scheme

The items in each ranking question were presented in a randomized order to each survey participant.

Section 3: Security Awareness

Final section was a quiz of multiple choice questions, focusing on mobile security awareness. These questions were set using common misconceptions about mobile phone security. The score from these questions will be used as a measure for the security awareness of the respondents.

The complete set of questions used in this survey can be found in the full paper [29].

Survey responses were gathered through online portals SurveyMonkey and Cint in January 2018. The respondents were paid based on the standard rates in the online platforms. The impact of security awareness on different categories was tested using p-values.

3.2 *Results*

The responses from the survey were analyzed in order to understand the perception of the users towards CA. Different demographic and other factors influencing these perceptions is also studied here.

3.2.1 Response Demographics

In total 695 people responded to the survey, out of which 494 (71%) were complete responses. These 494 was used for the rest of the analysis. The age demographics of the respondents is shown in Fig. 3.

Out of the respondents 328 (66%) were female and 163 (33%) were male, 3 people (1%) preferred not to disclose their gender.

Majority of the respondents (90%) were from the USA.

In total 6 (1.2%) respondents did not attend school, 223 (45%) respondents had completed high school education, 175 (35%) has completed university degree and 90 (18%) has a higher degree, MSc or PhD.

Out of the respondents, 82 (17%) identified themselves as working in Computer & technology related fields.

Fig. 3 Age demographics of the respondents

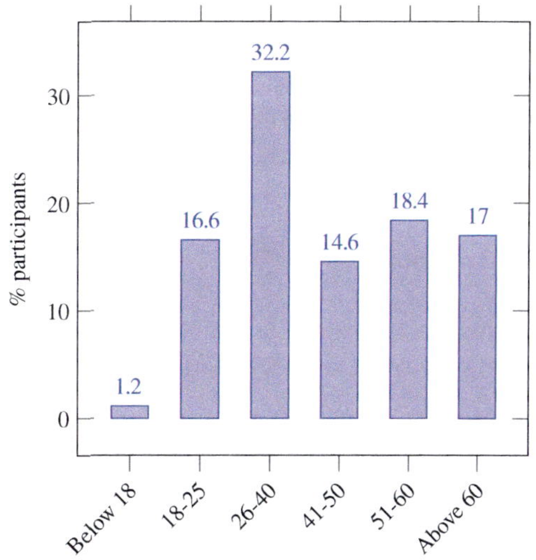

3.2.2 Security Awareness

The respondents were surveyed with 5 questions and the scored out of 5. The mean score for security awareness quiz is $\mu = 2.66$ and variance is $\sigma = 1.26$ These scored were used to categorize the users into two categories,

1. High Security Awareness (HSA): Scoring 3 and above on the quiz (294 - 60%) $\mu = 3.5340$ and $\sigma = 0.6690$
2. Low Security Awareness (LSA): Scoring below 3 on the quiz (199 - 40%) $\mu = 1.3768$ and $\sigma = 0.7062$

These scores were used to analyze the hypotheses H1 - H5 as follows,

- *H1. Gender impact on security awareness*
  Males have above average security awareness ($P = 0.04975$). Even though mean score for females was below average, there was no statistically significant evidence that females were less security aware. ($P = 0.07706$)
- *H2. Age impact on security awareness*
  Our survey shows that people got are more security aware as they got older. This finding went against common belief that younger people are more technology aware.
- *H3. Impact of occupation*
  There was no impact from the occupation being Computer related and non-Computer related towards the security awareness. ($P = 0.2107$ & $P = 0.2289$)
- *H4. Impact of education level*
  Security awareness increases with education level by looking at these results.

For a more detailed analysis about these conclusions with statistical significance value calculations please refer to the paper [29].

3.2.3 Mobile Device OS

Current mobile phone usage of the respondents was analyzed and shown in Fig. 4. The figure show the HSA and LSA respondents for each device type.

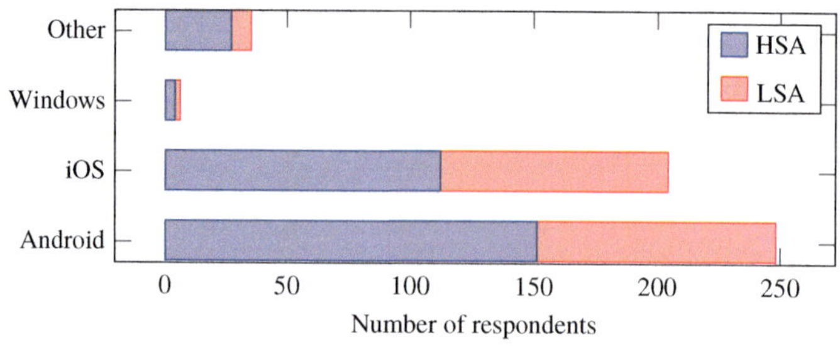

Fig. 4 Device OS and security awareness

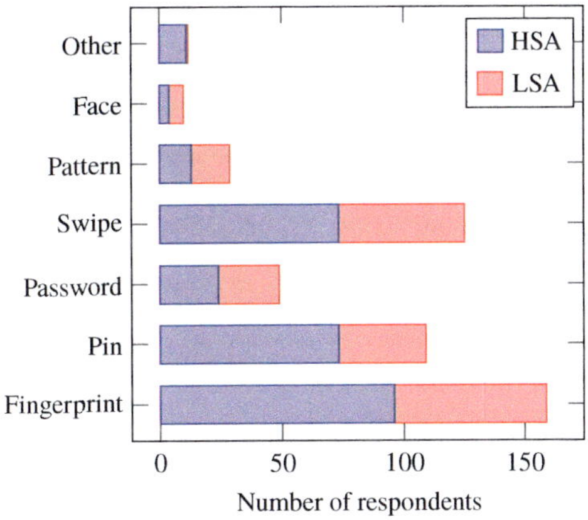

Fig. 5 Current screen unlock method

H5. Impact of device OS to security awareness

Looking at the individual scores in each category of mobile device we can conclude the following with statistical significance,

- People using iOS phones have below average security awareness. ($P = 0.0449$)
- People using Android phones do not show statistically significant differences of security awareness. ($P = 0.1256$)
- People using other phones have above average security awareness. ($P = 0.0194$)

Benenson et al. in [2] attempted to measure the security awareness by the presence or absence of a virus guard in the users mobile device. However having a virus guard is not a direct measure of security awareness. Here we use the quiz score which provides a direct measure of security awareness.

3.2.4 Current Screen Lock Method *(H6)*

Different screen lock methods used by the respondents is shown in Fig. 5.

The most popular unlock method is fingerprint with the others follow in order of Swipe, Pin, Password, Pattern, Other and FaceID.

It is important to note that the second most popular authentication method is Swipe (i.e. not having any authentication).

Convenience of screen lock method

A 5 point Likert scale was used to evaluate the different unlock methods used based on their convenience. The results of this question is shown in Fig. 6.

Fingerprint and swipe are the most convenient unlock methods followed by face. PIN and password have the highest inconvenient ratings, and it is interesting that none of the pattern users marked it as inconvenient.

The respondents were asked to highlight issues and inconveniences with possible improvements to their current authentication methods. Most common issues included the authentication method being slow and not registering in one try. The

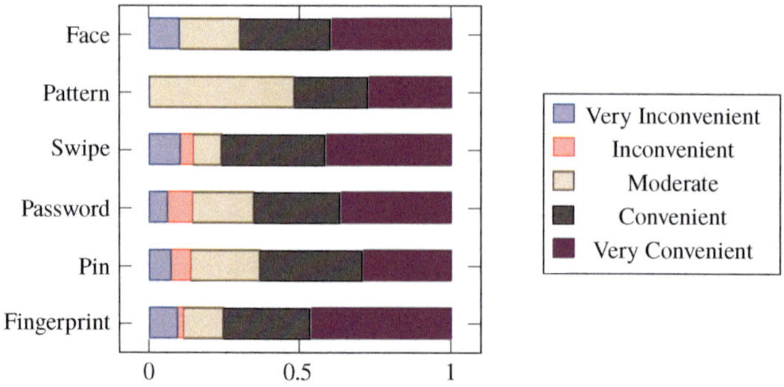

Fig. 6 Convenience of current authentication method

Fig. 7 Requirement of levels of security

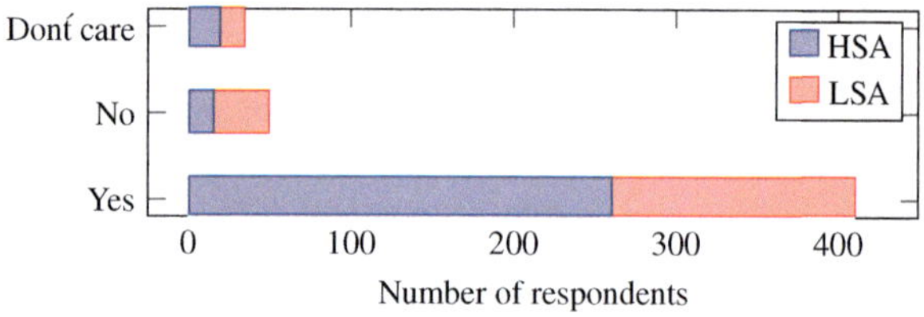

comments show that, even with the widespread usage of current unlock methods there are many issues with reliability and convenience in each of those.

3.2.5 Levels of Security *(H7)*

The respondents were asked if different apps require different security levels, the result of this question is shown in Fig. 7.

An overwhelming 83.16% agree that different apps require different security levels.

H8. The perceived requirement of different levels of security depends on security awareness

- The people who said "No" have below average security awareness ($\mu = 1.1938$, P-value 0.00002)
- The people who said "Yes" have above average security awareness. ($P = 0.0212$).

With this evidence it is clear that varying security levels is a requirement evident to end users. However existing mobile phone operating systems do not provide a multilevel security scheme.

The respondents were asked to rank different applications in the order of security requirement. This ranking activity resulted in a security requirement score for each app shown in the Fig. 8.

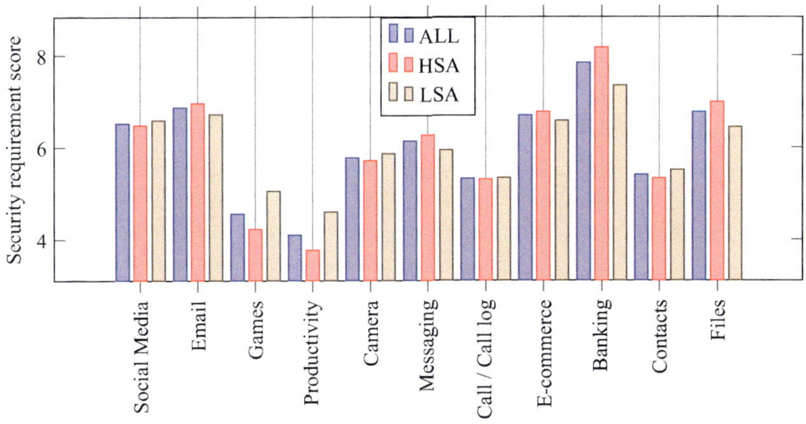

Fig. 8 Security requirement for different mobile applications/features

The averages from all responses, HSA and LSA are given separately. There is a slight difference between the ordering given by HSA and LSA. It is clear that both HSA and LSA think Banking apps require the highest level of security.

- HSA have rated Files/Storage higher than LSA responses
- LSA have rated social media and Messaging(SMS) higher than HSA responses

It is evident that security requirements differ from person to person and the personal preference might be affected by factors like security awareness.

3.2.6 Biometrics

The respondents were asked to rank different biometrics and traditional secrets-based authentication methods from the least secure to the most secure. The score obtained by this ranking is given in Fig. 9.

The overall ranking of the biometrics from most secure to least secure is as follows, (1) Iris Scan, (2) Fingerprint, (3) Face, (4) Voice, (5) Password, (6) PIN and (7) Pattern.

The only difference in the ranking between HSA and LSA was, LSA ranked Fingerprint above Iris. An interesting observation is that all biometrics-based authentication methods were ranked above secrets-based authentication methods.

3.2.7 Continuous Authentication

Willingness to use ***(H9)***
Based on the demonstration video and descriptions of CA the respondents were asked about their willingness to use it. The responses are evaluated in Fig. 10.

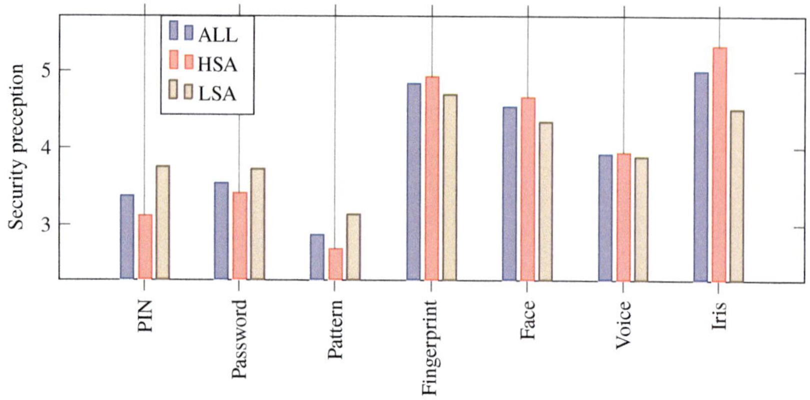

Fig. 9 Security perception for different authentication methods

Fig. 10 Willingness to use CA

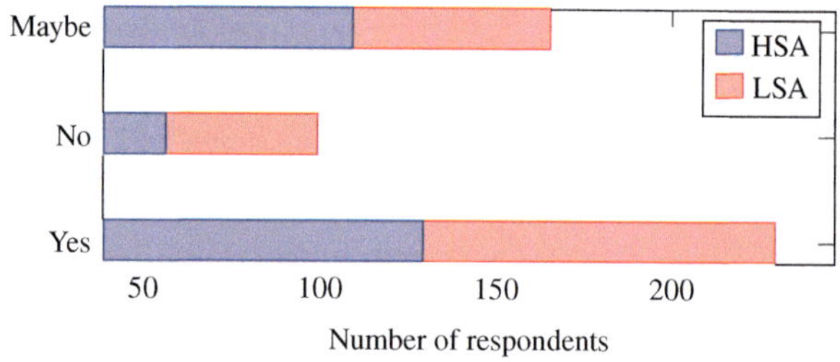

46.6% would use CA, while 20% will not, the other 33.4% responded with maybe.

H10. Security awareness has an impact on the willingness to use Continuous Authentication

- Respondents who said "Maybe" are people with higher security awareness. ($P = 0.0132$)
- People who said "Yes" are people with lower security awareness. ($P = 0.0393$)

This highlights how lower security aware people are more trusting towards new technologies and are willing to try without further investigation. Whereas people who said "Maybe" are more security aware and needs more proof and assurance about the new authentication system.

Most of the people who were willing to use CA cited the convenience, security and safety provided and ease of sharing the device with friends or family.

The people who are not willing to use CA say they do not think it is secure and they do not trust it. Some people find the concept too complicated and they feel the current methods are sufficient. Another issue is the worry about biometrics, "*I don't want my phone constantly checking biometrics*."

Finally the people unsure of CA who responded "Maybe" said, they need more information before deciding. These users are more security aware and they needed

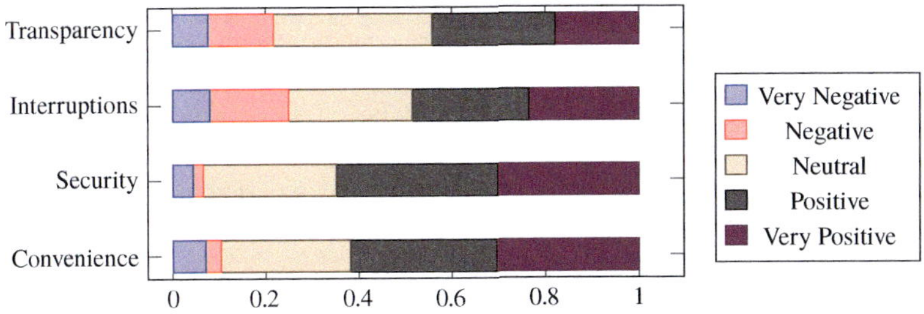

Fig. 11 Perceptions toward Continuous Authentication

to try it out and see how secure it is. Another common question raised is about the power consumption, "*it seems to me that continuously monitoring biometric signals will consume a lot of battery.*"

Looking at these responses we can see that, in order to propose a CA system for the masses, its essential to clearly show how it is more secure, how the collected biometric data will be handled and ensure that there is less overhead in terms of resource consumption like power.

Perceptions toward CA *(H11)*

5 point Likert scale questions were employed to evaluate user perceptions across 4 different aspects of CA. Figure 11 shows the results for these four, Convenience, Security, Interruptions and Transparency.

Over 60% found CA very convenient or convenient. Only 10% found it Inconvenient.

Over 65% found CA more secure than traditional methods.

The additional interruptions that might be introduced due to CA were seen as annoying by over 20% of the respondents. Similarly, the transparency of CA was seens as a negative by roughly 20% of the respondents. Therefore a feedback mechanism to ensure constant feedback to the user regarding the authentication process is essential.

In order to understand how different factors affect the perceptions toward CA, chi-square tests was run on all the Likert scale questions considering Security awareness, Gender and Current mobile device. The complete results of these tests are presented in our paper [29].

In order to identify what aspects users consider when selecting an authentication scheme, they were asked to rank 5 aspects from the least important to the most important. The results from this ranking was consistent with HSA and LSA respondents. The final ranking of the factors from most important to least important was as follows, (1) Security, (2) Ease of use, (3) Time taken to unlock, (4) Power Consumption and (5) Ease of setup.

3.3 Discussion

The respondents gave suggestions and what they felt about CA in an open-ended question. Majority of the users (160 responses) gave positive responses and showed willingness to adopt CA if it is deployed. Some of the suggestions and desired features are highlighted with following quotes,

- Configurable security levels for user profiles
- Configurable security levels for apps
- To have a toggle switch to easily turn off CA for instances where free device sharing is needed,
- Have a feedback mechanism to show which user the phone identifies currently.
- Ease of use and setup,
- To ensure its fast and reliable

These suggestions could be used as key considerations when developing a Continuous Authentication system for widespread use on mobile devices.

3.4 Limitations

We identified the following limitations of the survey design and response collection process, which might have an adverse effect on the statistical results generated.

- A major limitation of the survey is that the respondents did not actually use a Continuous Authentication system themselves; instead, they watched a scenario-based video that explained the concept. This could lead to respondents being optimistic about the pitfalls, such as the inconvenience of frequent false rejects, in a Continuous Authentication scheme.
- Respondents could be more technology-aware than the general population since they were recruited from an online portal. Therefore the security awareness, education level and perceptions analyzed might not be representative of the general population.
- Respondents' demographics might not properly represent the total smart phone user base. For example, 66% of the respondents were females, while only 33% were males. This might affect how well one can generalize our survey results to a target market.

4 Resource Consumption

4.1 *Methodology*

In this section, the goal is to understand how accurate a biometric modality (say, fingerprint) is, in relation to how much computation resources it consumes. Intuitively, the more resources a modality can use (e.g. extracting more features, combining different classifiers), the more accurate we expect its authentication to be. Conversely, if a modality uses very little computation power (e.g. by random guessing), then its accuracy will be low.

On the one hand, the accuracy of a given biometric modality depends on two main factors:

1. The *Inherent Uniqueness* of the biometric modality. It is well known that some modalities, like fingerprint and iris, are more unique (ie. more discriminative) than face or voice or gait.
2. The *Discriminating Power* of the classification algorithm, which depends on the features extracted, and the choice of classifier.

On the other hand, the resources consumed by the biometric modality also depend on two factors:

1. *Acquisition overhead*: This refers to the energy consumed by the sensor(s) that acquire the said biometric modality. Examples: for voice, the sensor is the microphone (which may be turned off when not in use); for gait, the sensor is the accelerometer and gyroscope (which are usually turned on all the time).
2. *Algorithm consumption*: This refers to the energy and memory used by the algorithm, which in turn vary according to the features and classifier used.

We will be measuring all these factors to derive the Resource Profile Curves for each biometric modality being considered.

4.1.1 Biometric Algorithms

In order for a biometric modality to be suitable for use in a CA scenario, the signal acquisition should be passive and non-intrusive. Therefore some physiological biometrics like fingerprint and iris are not suitable as they are implemented today because they require users active cooperation to capture. In our study, we have selected,

1. Face
2. Voice
3. Touch Screen Gestures
4. GAIT
5. Soft/Geometric Face Features

Table 1 Datasets used

Dataset	Modalities	Identities	Sessions
MOBIO [23]	Video, Audio	152 total	2 sessions, 6 rounds each
Touchalatics [16]	Smartphone touch	40 people	1
HuGaDB [6]	Accelerometer data	18 people	Variable number

as the biometric modalities which allow the acquisition to be done transparently.

In order to characterize the resource consumption vs utility of each of these biometric modalities, some of the popular implementations for these biometric modalities were selected and implemented.

Multiple variations of algorithms for these biometric modalities were analyzed. Multiple combinations of features, classifiers along with variations of different parameters were used to get different configurations of algorithms for each modality. More details of the algorithms are provided in the paper [30]. These different configurations will later be profiled in terms of energy consumption and memory consumption to get their Resource Profile Curves.

4.1.2 Datasets

In order to test all of the different algorithms on a fair grounds we needed a dataset which provides input for all 5 modalities. Since there is no existing dataset that satisfies the requirement, we combined 3 different datasets as shown in Table 1 to create virtual identities.

A key consideration when selecting the datasets was that, they have to emulate the realistic complexity of biometric modalities captured within a mobile environment for CA. Therefore all the datasets were selected to be captured in mobile phones and in usual usage scenarios.

In order to keep the datasets in a similar complexity, we ensured the number of different identities was kept similar. To achieve this we used the IDIAP collection (26 identities) data on MOBIO and entire datasets of Touchalatics and HuGaDB(40 and 18 identities).

4.1.3 Resource Profile Curves (RPC)

The objective of the study is to plot a curve for resources consumed (horizontal axis) versus the utility provided by each biometric modality (vertical axis). Here the utility for any biometric is the level of security that modality is able to provide. The level of security will be measured by its classification accuracy.

The ideal RPC would thus be an inverted "L" shape, where the perfect accuracy is achieved with the minimum amount of resources. The worst case RPC would be a horizontal line on the x-axis where, regardless of the resources consumed, the accuracy remains at a minimum. However, in reality, the worst case is lower-bounded by the RandomGuesser – the algorithm that randomly accepts or rejects the user.

Each of the algorithm configurations can be plotted with respect to accuracy vs resource consumption in a scatter plot. Let,

$$S = \{(\text{resource, accuracy})\text{pairsforeachmodality}\} \quad (1)$$

Using the points in S an RPC needs to be generated. In order to generate this the following observations were used,

- The least energy consuming algorithm will be a random guess which will also give the lowest accuracy
- For any limit in available resource level, the algorithm which provides the best accuracy for a lower resource consumption level will be selected

Therefore, the Resource Profile Curve will be lower bounded by the random guessing algorithm. Any new points in the RPC should be to the right and above this random point. Therefore the RPC will be a monotonically increasing curve.

To generate the RPC the critical points for each modality will be selected using the method shown in the Algorithm 1. Here $p.RC$, $q.RC$ refers to the resource consumption and $p.acc$, $q.acc$ refers to the accuracy of p and q.

The Resource Profile Curve will be drawn using the pairwise linear curve on the critical points in Sc generated as shown in the algorithm. Following the same method, two Resource Profile Curves were generated,

- Accuracy vs Energy consumption (EC) Profile
- Accuracy vs Memory consumption (MC) Profile

Algorithm 1: Isolating critical points

Data: S = {Points in the scatter plot}
Result: Sc = {Critical points}
Let Sc = { };
foreach $p \in S$ **do**
    $critical = True$
    foreach $q \in S \mid q.RC \leq p.RC$ **do**
        if $p.acc \leq q.acc$ **then** $critical = False$
    end
    if $critical = True$ **then** $Sc.add(p)$
end

4.1.4 Measuring Energy Consumption (EC)

The overall energy consumption for an authentication task can be analyzed in two parts,

1. EC of the algorithm to perform authentication
2. EC of the sensor to acquire the biometric signal

Energy consumption (EC) for algorithm
Time consumed for recognition by each algorithm configuration was used as a proxy for the EC. The main assumption was that the EC by the mobile device will be proportional to the execution time for each algorithm,

$$Energy \propto Time \tag{2}$$

$$Energy = k \times Time \tag{3}$$

Here k is the constant of proportionality.

To calculate k, a simple algorithm with a set number of calculations was executed in the PC as well as the Android environment. The runtime for one instance of the algorithm will then be measured and the rate of discharge of the phone battery will also be measured. These two values will be used in the Eq. 3 to calculate k.

Energy consumption for acquisition
An Android application was developed to continuously log the battery level over time at a constant interval. The discharge rate was calculated for the following: idle, capturing face image every 5 s, capturing a voice clip every 5 s, logging accelerometer data, logging touch screen data.

In order to isolate the energy discharge rates for biometric signal inputs(RD_{sensor}) the rate of discharge of idle state (RD_{idle}) was deducted from the total rate of discharge ($RD_{measured}$) as shown in Eq. 4

$$RD_{sensor} = RD_{measured} - RD_{idle} \tag{4}$$

The energy consumption ($Energy_{algo}$) for a given algorithm can be then calculated as follows,

$$Energy_{algo} = k \times Time_{algo} + R_{sensing} \tag{5}$$

The time consumption was measured on a core i7-6700 3.4 GHz CPU with 8 GB of RAM. Mobile battery discharge rates were measured using an LG V10 android device.

The main focus here, is on the time/energy consumed in the test environment. The time consumption for training the models is not considered because in practice training will be done only once, when registering the user of a smartphone. However, the test scenario has to be run on the mobile devices continuously to achieve CA.

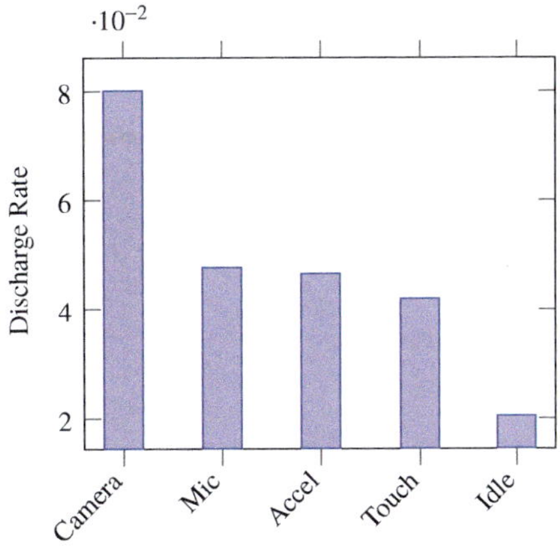

Fig. 12 Energy consumption of different sensors

4.1.5 Measuring Memory Consumption

The total sizes of feature extraction models (where needed) and classification models were added up for each configuration to measure the memory consumption of each method.

4.2 Results

The rate of discharge results is shown in Fig. 12. The highest energy consuming sensor is the camera and the lowest energy consuming sensor is the touch screen sensor.

4.2.1 Calculating Energy Consumption

To calculate the constant of proportionality (k) a simple number addition algorithm was implemented in both computer and mobile platforms and the time and energy discharge rates were measured and the value for k was calculated using the Eq. 3. The value for k was a very high value (≥ 150). Therefore, based on Eq. 5 the time consumption of the algorithms would dictate the behavior of the curve and hence, the energy consumption of sensing action has a negligible effect.

4.2.2 Accuracy vs Energy Consumption Profile

Figure 13 shows the energy consumption vs accuracy curve. The x-axis has been log scaled in order to expand the smaller values and compress the larger values.

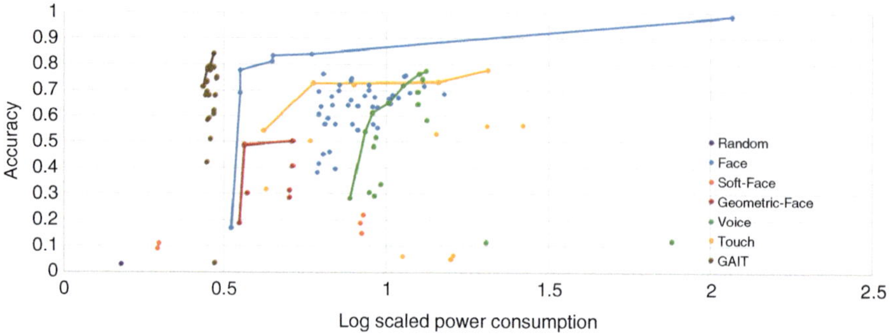

Fig. 13 Energy profile for different biometrics (best viewed in color)

All the curves start with a low energy consumption, low accuracy state and they provide higher accuracy with increasing energy consumption. The curves eventually flatten out as energy consumption increases, this shows diminishing returns as the amount of energy consumed is increased.

It can be observed that GAIT outperforms most biometrics in low energy consumption, however, Face biometric outperforms all of the other modalities for highest accuracy achieved. The highest performing algorithm configuration for Face (VGG) consumes roughly 3 times the energy of the 2nd best Face-based user recognition algorithm.

The random point shown in the graph is a baseline for the lowest energy consumption and lowest accuracy, it can be observed that by increasing a very small amount of energy we can achieve a slight increase in accuracy by using soft biometric traits (skin color).

4.2.3 Accuracy vs Memory Consumption Profile

Figure 14 shows the memory consumption vs accuracy curve, similar to the previous graph this graph's x-axis has also been log-scaled.

When considering memory constraints there is no clear leader. We can achieve a better than random accuracy without having to save a trained model by using soft feature-based methods. Looking at the curves we can see that voice and GAIT performs best for lower memory values and with larger model sizes face biometrics outperforms the rest.

We will see practical usages of this curve in Sect. 5.

4.3 Discussion

Comparing Figs. 13 and 14, complex decisions can be made by a CA system. Depending on the available memory and battery level an intelligent CA system

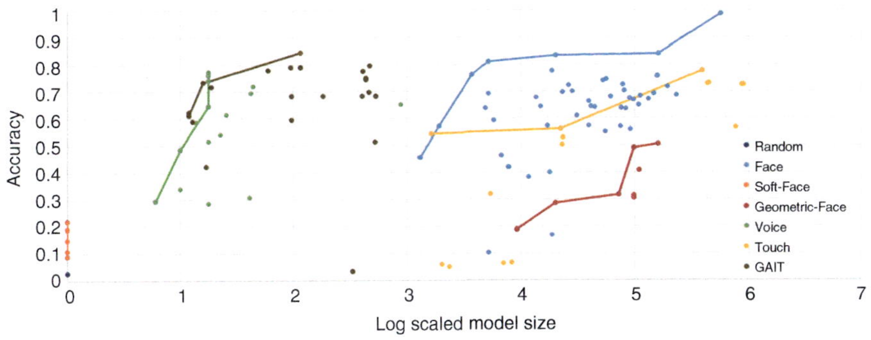

Fig. 14 Memory profile for different biometrics (best viewed in color)

should be able to provide the maximum possible security by using these RPCs. We will illustrate the use-cases highlighted in Sect. 1 with the generated Resource Profile Curves here,

Security-First

Higher security requiring applications like Banking would require a high-security level (say, accuracy levels over 0.9). Using the RPCs in Figs. 13 & 14 it is clear that, in order to achieve this level of accuracy the CA system can enforce the use of Face biometric. If unable to capture Face passively the CA system could prompt the user to explicitly provide an authentication before allowing access.

When using an application like YouTube the security requirement is comparatively lower. In a scenario like this, according to the RPC in Fig. 13 the CA system can limit power consumption by using GAIT or Soft-Face biometrics. The CA system can limit the memory consumption by using RPC in Fig. 14 to select Voice or GAIT.

Resource-First

Modern smartphones allow the user to select an energy saver mode, which would activate when the battery level of the device drops below a specified level. In a scenario like this, a CA system can operate in a lower region of the x-axis in the energy profile curve in Fig. 13.

A mobile device has a limited amount of free memory. If the available memory is low, CA system can use the memory profile curve in Fig. 14 to operate in a lower region of the x-axis. By choosing modalities like Soft-Face, GAIT and Voice the CA system can minimize the use of memory.

It is important to note that the RPCs shown here are for unimodal systems. Multiple points in these RPCs could be fused together to achieve higher levels of accuracy at higher resource consumption levels. There is also a trade-off between memory and energy which can be taken advantage of based on the resource limitations.

Context based

Based on the current context the mobile device is being used, the availability of the modalities will change. Following two examples illustrate these scenarios.

1. Walking while answering a call: In this scenario only voice and GAIT modalities will be available.
2. Sitting down, scrolling through an article: In this scenario only Face and Touch modalities will be available

In any of these scenarios, using the RPC, a CA system can find comparable algorithms for the available modalities. The modality selection can be based upon the security required. If the accuracy required is around 0.7, by looking at Fig. 13 we can see that there are comparable algorithms for GAIT, Face, Touch and Voice for this accuracy level.

For an example, when trying to select a biometric modality for CA when face and GAIT are unavailable (due to low light, sensor occlusion in a stationary use-case) the choice will depend on the level of accuracy needed. If the level of accuracy needed is around 0.3, the best alternative would be Soft-Face; if the accuracy requirement is around 0.6, the best alternative would be Touch and if the accuracy requirement is over 0.75 the only option would be Voice. This illustrates how the RPC enables smart choices for a CA system.

Even when a biometric modality is available, depending on the usage scenario the suitability of the biometric might vary. For an example, voice biometric could be appropriate in an indoor quiet environment, however it might be harder to use in a noisy outdoor environment. A fusion method taking these acquisition quality differences into account was proposed by Sivasankaran et al. [34].

4.4 Limitations

We identified the following limitations in the generated RPCs,

- The complexity of the datasets can affect the measurement of the Resource Profile Curves. To minimize the impact of this we chose the datasets to be comparable to actual usage in CA for mobile devices. For a given algorithm, the energy consumption will not vary based upon the dataset, however, the accuracy levels will vary based on the dataset. Therefore each modality can be represented by a band of values more completely than the current curves.
- As technology improves these curves will keep changing. However, it is clear that the curves can only keep moving up (achieving higher accuracy) and left (consuming lower resources). Therefore we can view these curves as a snapshot view of the biometric modalities. For an example, by looking at the Face RPCs in both Figs. 13 and 14 we can see that there is potential to try and reduce the memory consumed by Face-based authentication algorithms.

5 Conclusions and Future Work

We draw statistically significant conclusions about perception of biometrics and other authentication methods which shows how the perceptions have shifted over the years. It is clear that today's society is more accepting towards biometrics based methods compared to 20 years ago.

We see a very positive response towards multi-level security schemes for mobile devices which can be a welcome addition to the current main stream mobile operating systems.

Perceptions towards Continuous Authentication systems show that most people find it very useful and are willing to adopt CA. We also highlight key concerns and features that would need to be addressed in a Continuous Authentication system for mobile devices.

The two sets of RPCs generated in this work provides a new perspective towards evaluating the suitability of biometrics for constrained environments like mobile devices. It is important to note that these curves will keep shifting as the algorithms and hardware improve.

One of the future work is to extend the curves into bands of values by varying the complexity of the datasets as discussed in Sect. 5.

Another target is to use these Resource Profile Curves in an intelligent decision-making engine to dynamically switch between biometric modalities depending on their availability, resource availability and security requirement.

Epilog: Bob should use Continuous Authentication on his phone. Now, whenever bob picks up his phone, it will automatically detect the motion and capture biometric data like face image or voice sample, and use biometrics to automatically unlock the phone. The CA will keep on checking to see if Bob is using the phone. When Bob passes his phone to Alice, the phone detects that the user is no longer Bob. When Alice tries to open Facebook, CA will not allow Alice to proceed into the personal details of Bob. With the new CA system Bob does not worry about sharing his phone freely. The CA system uses RPCs to conserve power consumption, and therefore, Bob does not have to worry about phone battery drain either.

References

1. Ben-Asher N, Kirschnick N, Sieger H, Meyer J, Ben-Oved A, Möller S (2011) On the need for different security methods on mobile phones. In: Proceedings of the 13th international conference on human computer interaction with mobile devices and services, MobileHCI '11. ACM, pp 465–473, New York. ISBN 978-1-4503-0541-9. https://doi.org/10.1145/2037373.2037442

2. Benenson Z, Gassmann F, Reinfelder L (2013) Android and ios users' differences concerning security and privacy. In: CHI '13 extended abstracts on human factors in computing systems, CHI EA '13. ACM, pp 817–822, New York. ISBN 978-1-4503-1952-2. https://doi.org/10.1145/2468356.2468502
3. Bours P (2012) Continuous keystroke dynamics: a different perspective towards biometric evaluation. Inf Secur Tech Rep 17(1):36–43. ISSN 1363-4127. https://doi.org/10.1016/j.istr.2012.02.001. http://www.sciencedirect.com/science/article/pii/S1363412712000027. Human Factors and Bio-metrics
4. Braz C, Robert J-M (2006) Security and usability: the case of the user authentication methods. In: Proceedings of the 18th conference on l'interaction homme-machine, IHM '06. ACM, pp 199–203, New York. ISBN 1-59593-350-6. https://doi.org/10.1145/1132736.1132768
5. Carroll A, Heiser G, et al (2010) An analysis of power consumption in a smartphone. In: USENIX annual technical conference, vol 14, Boston, pp 21–21
6. Chereshnev R, Kert'esz-Farkas A (2017) Hugadb: human gait database for activity recognition from wearable inertial sensor networks. In: International conference on analysis of images, social networks and texts. Springer, pp 131–141
7. Clarke N, Furnell S (2005) Authentication of users on mobile telephones, a survey of attitudes and practices. Comput Secur 24(7):519–527. ISSN 0167-4048. https://doi.org/10.1016/j.cose.2005.08.003. http://www.sciencedirect.com/science/article/pii/S0167404805001446
8. Clarke N, Karatzouni S, Furnell S (2009) Flexible and transparent user authentication for mobile devices. In: Gritzalis D, Lopez J (eds) Emerging challenges for security, privacy and trust. Springer, Berlin/Heidelberg, pp 1–12. ISBN 978-3-642-01244-0
9. Clarke N, Karatzouni S, Furnell S (2009) Flexible and transparent user authentication for mobile devices. In: IFIP international information security conference. Springer, pp 1–12
10. Crawford H, Renaud K, Storer T (2013) A framework for continuous, transparent mobile device authentication. Comput Secur 39:127–136. ISSN 0167-4048. https://doi.org/10.1016/j.cose.2013.05.005. http://www.sciencedirect.com/science/article/pii/S0167404813000886
11. Creus BG, Kuulusa M (2007) Optimizing mobile software with built-in power profiling. In: Mobile phone programming. Springer, pp 449–462. https://link.springer.com/chapter/10.1007/978-1-4020-5969-8_25
12. Deane F, Barrelle K, Henderson R, Mahar D (1995) Perceived acceptability of biometric security systems. Comput Secur 14(3):225–231. ISSN 0167-4048. https://doi.org/10.1016/0167-4048(95)00005-S. http://www.sciencedirect.com/science/article/pii/016740489500005S
13. Falaki H, Mahajan R, Estrin D (2011) Systemsens: a tool for monitoring usage in smartphone research deployments. In: Proceedings of the sixth international workshop on MobiArch. ACM, pp 25–30
14. Feng T, Liu Z, Kwon KA, Shi W, Carbunar B, Jiang Y, Nguyen N (2012) Continuous mobile authentication using touchscreen gestures. In: 2012 IEEE conference on technologies for homeland security (HST), pp 451–456. https://doi.org/10.1109/THS.2012.6459891
15. Frank M, Biedert R, Ma E, Martinovic I, Song D (2013) Touchalytics: on the applicability of touchscreen input as a behavioral biometric for continuous authentication. IEEE Trans Inf Forensics Secur 8(1):136–148. ISSN 1556-6013. https://doi.org/10.1109/TIFS.2012.2225048
16. Frank M, Biedert R, Ma E, Martinovic I, Song D (2013) Touchalytics: on the applicability of touchscreen input as a behavioral biometric for continuous authentication. IEEE Trans Inf Forensics Secur 8(1):136–148, 1. ISSN 1556-6013. https://doi.org/10.1109/TIFS.2012.2225048
17. Guerra-Casanova J, Ríos-Sànchez B, Viana-Matesanz M, Bailador G, Sànchez-Àvila C, Giles MJMD (2016) Comfort and security perception of biometrics in mobile phones with widespread sensors. In: 2016 IEEE 35th symposium on reliable distributed systems workshops (SRDSW), Sept 2016, pp 13–18. https://doi.org/10.1109/SRDSW.2016.13
18. Karlson AK, Brush AB, Schechter S (2009) Can I borrow your phone? Understanding concerns when sharing mobile phones. In: Proceedings of the SIGCHI conference on human factors in computing systems, CHI '09. ACM, New York, pp 1647–1650. ISBN 978-1-60558-246-7. https://doi.org/10.1145/1518701.1518953

19. Khan H, Hengartner U, Vogel D (2015) Usability and security perceptions of implicit authentication: convenient, secure, sometimes annoying. In: Eleventh symposium on usable privacy and security (SOUPS 2015). USENIX Association, Ottawa, pp 225–239. ISBN 978-1-931971-249. https://www.usenix.org/conference/soups2015/proceedings/presentation/khan
20. Khan H, Hengartner U, Vogel D (2015) Usability and security perceptions of implicit authentication: convenient, secure, sometimes annoying. In: SOUPS, pp 225–239
21. Kumar R, Phoha VV, Serwadda A (2016) Continuous authentication of smartphone users by fusing typing, swiping, and phone movement patterns. In: 2016 IEEE 8th international conference on biometrics theory, applications and systems (BTAS), Sept 2016, pp 1–8. https://doi.org/10.1109/BTAS.2016.7791164
22. Matthews T, Liao K, Turner A, Berkovich M, Reeder R, Consolvo S (2016) "she'll just grab any device that's closer": a study of everyday device & account sharing in households. In: Proceedings of the 2016 CHI conference on human factors in computing systems, CHI '16. ACM, New York, pp 5921–5932. ISBN 978-1-4503-3362-7. https://doi.org/10.1145/2858036.2858051
23. McCool C, Marcel S, Hadid A, Pietikainen M, Matejka P, Cernocky J, Poh N, Kittler J, Larcher A, Levy C, Matrouf D, Bonastre J-F, Tresadern P, Cootes T (2012) Bi-modal person recognition on a mobile phone: using mobile phone data. In: IEEE ICME workshop on hot topics in mobile multimedia, July 2012
24. Micallef N, Just M, Baillie L, Halvey M, Kayacik HG (2015) Why aren't users using protection? investigating the usability of smartphone locking. In: Proceedings of the 17th international conference on human-computer interaction with mobile devices and services, MobileHCI '15. ACM, New York, pp 284–294. ISBN 978-1-4503-3652-9. https://doi.org/10.1145/2785830.2785835
25. Mosenia A, Sur-Kolay S, Raghunathan A, Jha NK (2017) Caba: continuous authentication based on bioaura. IEEE Trans Comput 66(5):759–772. ISSN 0018-9340. https://doi.org/10.1109/TC.2016.2622262
26. Niinuma K, Park U, Jain AK (2010) Soft biometric traits for continuous user authentication. IEEE Trans Inf Forensics Secur 5(4):771–780. ISSN 1556-6013. https://doi.org/10.1109/TIFS.2010.2075927
27. Papavasileiou I, Smith S, Bi J, Han S (2017) Gait-based continuous authentication using multimodal learning. In: 2017 IEEE/ACM international conference on connected health: applications, systems and engineering technologies (CHASE), July 2017, pp 290–291. https://doi.org/10.1109/CHASE.2017.107
28. Qian F, Wang Z, Gerber A, Mao Z, Sen S, Spatscheck O (2011) Profiling resource usage for mobile applications: a cross-layer approach. In: Proceedings of the 9th international conference on mobile systems, applications, and services, MobiSys '11. ACM, pp 321–334. ISBN 978-1-4503-0643-0. https://doi.org/10.1145/1999995.2000026
29. Rasnayaka S, Sim T (2018) Who wants continuous authentication on mobile devices? In: International conference on biometrics techniques applications and systems
30. Rasnayaka S, Saha S, Sim T (2019) Making the most of what you have! profiling biometric authentication on mobile devices. In: International conference on biometrics
31. Sawaya Y, Sharif M, Christin N, Kubota A, Nakarai A, Yamada A (2017) Self-confidence trumps knowledge: a cross-cultural study of security behavior. In: Proceedings of the 2017 CHI conference on human factors in computing systems, CHI '17. ACM, New York, pp 2202–2214. ISBN 978-1-4503-4655-9. https://doi.org/10.1145/3025453.3025926
32. Sieger KNMH (2010) User preferences for biometric authentication methods and graded security on mobile phones. In: Symposium on usability, privacy, and security (SOUPS) 2010
33. Sim T, Zhang S, Janakiraman R, Kumar S (2007) Continuous verification using multimodal biometrics. IEEE Trans Pattern Anal Mach Intell 29(4):687–700. ISSN 0162-8828. https://doi.org/10.1109/TPAMI.2007.1010
34. Sivasankaran D, Ragab MS, Sim T, Zick Y (2018) Context-aware fusion for continuous biometric authentication. In: International conference on biometrics

35. Srivastava S, Sudhish PS (2016) Continuous multi-biometric user authentication fusion of face recognition and keystoke dynamics. In: 2016 IEEE region 10 humanitarian technology conference (R10-HTC), Dec 2016, pp 1–7. https://doi.org/10.1109/R10-HTC.2016.7906823
36. Tiwari V, Malik S, Wolfe A (1994) Power analysis of embedded software: a first step towards software power minimization. IEEE Trans Very Large Scale Integr VLSI Syst 2(4):437–445
37. Traore I, Woungang I, Obaidat MS, Nakkabi Y, Lai I (2012) Combining mouse and keystroke dynamics biometrics for risk-based authentication in web environments. In: 2012 fourth international conference on digital home, Nov 2012, pp 138–145. https://doi.org/10.1109/ICDH.2012.59
38. Wagner DT, Rice A, Beresford AR (2013) Device analyzer: understanding smartphone usage. In: International conference on mobile and ubiquitous systems: computing, networking, and services. Springer, pp 195–208

9783030394882